ONE ~~WEE~~K LOAN

13 MA

Civil Engineering Claims

Third Edition

Vincent Powell-Smith
LLB(Hons), LLM, DLitt, FCIArb

Douglas Stephenson
BSc, CEng, FICE, FIStructE, FCIArb, MConsE

John Redmond
BA, FCIArb, Solicitor

Foreword by Lord Justice Otton

**Blackwell
Science**

© Douglas Stephenson and the estate of Vincent Powell-Smith, 1989, 1994; Douglas Stephenson, John Redmond and the estate of Vincent Powell-Smith, 1999

Blackwell Science Ltd
Editorial Offices:
Osney Mead, Oxford OX2 0EL
25 John Street, London WC1N 2BL
23 Ainslie Place, Edinburgh EH3 6AJ
350 Main Street, Malden
 MA 02148 5018, USA
54 University Street, Carlton
 Victoria 3053, Australia
10, rue Casimir Delavigne
 75006 Paris, France

Other Editorial Offices:

Blackwell Wissenschafts-Verlag GmbH
Kurfürstendamm 57
10707 Berlin, Germany

Blackwell Science KK
MG Kodenmacho Building
7–10 Kodenmacho Nihombashi
Chuo-ku, Tokyo 104, Japan

First edition 1989
Second edition 1994, reprinted 1997
Third edition published 1999

Set in 10.5/12.5 pt Palatino
by DP Photosetting, Aylesbury, Bucks
Printed and bound in Great Britain by
MPG Books Ltd, Bodmin, Cornwall

The Blackwell Science logo is a trade mark of Blackwell Science Ltd, registered at the United Kingdom Trade Marks Registry

DISTRIBUTORS

 Marston Book Services Ltd
 PO Box 269
 Abingdon
 Oxon OX14 4YN
 (*Orders:* Tel: 01235 465500
   Fax: 01235 465555)

USA
 Blackwell Science, Inc.
 Commerce Place
 350 Main Street
 Malden, MA 02148 5018
 (*Orders:* Tel: 800 759 6102
   781 388 8250
   Fax: 781 388 8255)

Canada
 Login Brothers Book Company
 324 Saulteaux Crescent
 Winnipeg, Manitoba R3J 3T2
 (*Orders:* Tel: 204 837-2987
   Fax: 204 837-3116)

Australia
 Blackwell Science Pty Ltd
 54 University Street
 Carlton, Victoria 3053
 (*Orders:* Tel: 03 9347 0300
   Fax: 03 9347 5001)

A catalogue record for this title is available from the British Library

ISBN 0-632-05197-3

Library of Congress
Cataloging-in-Publication Data
Powell-Smith, Vincent.
 Civil engineering claims/Vincent Powell-Smith, Douglas Stephenson, and John Redmond. – 3rd ed.
  p. cm.
 Includes index.
 ISBN 0-632-05197-3 (pb)
 1. Civil engineering contracts – Great Britain.
I. Stephenson, Douglas A. II. Redmond, John, B.A. III. Title.
KD1641.P672 1999
343.41'078624 – dc21   99-38807
                   CIP

For further information on
Blackwell Science, visit our website:
www.blackwell-science.com

CONTENTS

Contents

FOREWORD
by Lord Justice Otton

I still have on my shelves a copy of the First Edition of this work. I acquired it because I had come to the conclusion that judgments in construction cases (I admit, some of my own) were couched in too formulaic and arcane a style. They bore little resemblance to what happens in the real world when contractors, civil engineers and architects are either putting together a project, or executing it, and or disputing about it. If their decisions were to be acted upon it was incumbent upon judges (and arbitrators) when deciding or resolving disputes to provide a succinct factual matrix, to identify the issues between the parties, to evaluate the evidence (particularly of experts), to indicate the principles involved and to give their conclusions. My chosen path was not only to consider the tomes of learning (*Keating* and *Hudson* spring to mind) but to study the way in which professionals could talk to and instruct each other in plain English in a way which would be comprehensible to a wide spectrum of experience and try to express myself accordingly. The means were provided by my now battered copy and I hope that at least some of my judgments benefited from this approach.

From the perspective of the Court of Appeal the quality of both judgments and awards at first instance have improved considerably. In this regard the judges in the Technology and Construction Court in particular deserve recognition and praise.

Little did I realise that I should be invited by the publishers to contribute a Foreword to the Third Edition. I am privileged to do so. I congratulate the authors without reservation. Douglas Stephenson was one of the co-authors of the earlier edition and provides continuity of approach. John Redmond, a solicitor with unparalleled experience in this field (he is a former Chairman of the prestigious Society of Construction Law) brings a welcome innovative contribution.

Together they recognise that development of case law has (as predicted) maintained a momentum of its own. Judgments are available on the Internet within hours of hand-down in court, published in authoritative reports within days and their entrails pored over by experts and commentators within weeks. Busy practitioners need shielding from this indiscriminate bombardment. The authors have fastidiously incorporated only the bare minimum of case law by way of illustration and illumination. The curious can hunt elsewhere for more sustenance.

Similarly with the plethora of legislation which assails us from our domestic law and Europe. The authors have digested it and happily discriminated between the essential and peripheral. They take our hand and steer us through the maze.

Above all, there is a particular need for guidance on the new forms of contract and amendments to the old as they unremittingly encroach upon our lives. Practitioners had begun to live (albeit reluctantly) with the New Engineering Contract. Two years later the Engineering and Construction Contract arrived followed by the more welcome Design and Construct Form. Not surprisingly the fifth and sixth editions of the old ICE Conditions still found more favour than the novel. The Institute will no doubt continue to amend and up-date these forms to accommodate fresh legislation and decisions in the courts. It is in this field that the book comes into its own. The authors deal with these vexed and controversial matters in a concise, illuminating and pragmatic way. They also address the implications of the clumsily titled Housing Grants, Construction and Regeneration Act 1996, the Construction (Design and Management) Regulations 1994 and the Scheme for Construction Contracts Regulation which automatically triggers the still unproven adjudication of disputes. This particular topic will no doubt be the launching pad for the Fourth Edition!

The authors merit congratulations and praise. This is an invaluable, indeed essential part of every practitioner's library. It should be acquired by all engineers, contractors and sub-contractors, lawyers and (need I add?) judges concerned with civil engineering claims. My copy will find its place (in every sense) between my *Hudson* and *Keating*.

PREFACE

This book was first published in 1989 and was updated as a second edition in 1994. Since then developments in case law, on which Vincent Powell-Smith and I commented in the preface to the second edition, have continued at an ever-increasing pace; while the search for the holy grail in the form of a contract from which no dispute will ever arise was given new impetus by the publication, in 1993, of the New Engineering Contract. This was followed two years later by a second edition of that document, renamed the Engineering and Construction Contract. While it appears that as yet there is little if any reported case law arising from the form, the authors suspect that the absence of such authority results from reluctance of employers to use it rather than its success in eliminating disputes. Wisely, the parties to construction contracts appear generally to have preferred the trusted and respected Fifth and Sixth Editions of the ICE conditions, both derived from evolution since 1945, to a form having no such background, and relying mainly upon good-will for the avoidance of disputes.

In this edition we have continued to comment on the Fifth Edition of the ICE form in addition to the Sixth because, as a result of the government interference with construction contracts referred to below, the Institution of Civil Engineers has decided that it has no option but to continue updating the earlier form. Publication of the Seventh Edition is due in September 1999, and we have added a note to this preface covering the main differences between that edition and the updated version of the Sixth. Unless otherwise stated, references in this book to the ICE conditions may be taken to imply the Sixth Edition with amendments notified prior to the publication date of this book. Where desirable, we have emphasised words or phrases quoted from the various forms of contract by the use of italics; and in such cases the description '[authors' emphasis]' is to be implied (except in the NEC where 'terms identified in the Contract Data are in italics and defined terms have capital initials' (clause 11.1).

It has also been decided to add a chapter on the Engineering and

Construction Contract (or New Engineering Contract as it is still widely known). In addition the opportunity has been taken to add a further chapter on the widely-used Federation of Civil Engineering Contractors (now CECA) forms of sub-contract, from which important case law has arisen during the latter half of the 1990s.

A welcome addition to the ICE family of contracts, namely the Design and Construct Form, was first published in 1992 and was commented on in the second edition of this book. It was updated in 1995, and is based on the ICE Sixth Edition, incorporating the wealth of wisdom contained in that form. Surprisingly perhaps, it required only minor adjustments to incorporate design as an additional responsibility of the contractor. It appears to have generated little case law: but not, it is thought, for the reasons that apply in the case of the New Engineering Contract. We believe that the dearth of case law relating to the Design and Construct Form is attributable, in part at least, to the beneficial effects of integrating design with construction, resulting in collaboration of the parties responsible for those elements of the construction process instead of the mistrust, and sometimes antagonism, too often engendered by traditional procedures.

But perhaps the main innovation of the decade now drawing to its close is the degree of statutory intervention in construction contracts. Most, if not all, of the ICE forms of contract are affected by the Construction (Design and Management) Regulations 1994, and more particularly by the Housing Grants, Construction and Regeneration Act 1996. The latter has introduced a statutory requirement that construction contracts make provision for adjudication of disputes arising from non-payment; and, where the contract fails to make such provision, the Scheme for Construction Contracts (England and Wales) Regulations 1998 is deemed to be included in the contract.

Vincent Powell-Smith, who invited me to collaborate with him in writing the first and second editions of this book, died in 1997. He was a lawyer and a prolific author of enormous ability and great personality: a person with whom collaboration in book-writing was a pleasure. He will be sadly missed by his many friends and colleagues in the field of construction law.

Happily, after much deliberation with the publishers, we were able to find and to enlist the services of a construction contract lawyer of distinction, John Redmond, head of construction law at Laytons, Bristol. John had recently completed his term of office in the Chair of the Society of Construction Law, and was agreeable (or at least persuadable) to accept another challenge in the form of updating the predominately legal chapters of this volume. Working with him has been a pleasure; and his contribution to this third edition has been invaluable.

We have used the male pronoun throughout this book: not because the engineer and other professional people to whom we refer are necessarily male, but because there is as yet no simple and concise alternative. Happily the days when civil engineering was an exclusively male occupation have passed into history, and there is now, as in many other professions, a substantial and ever-increasing proportion of women, including many who have distinguished themselves in the ICE and other professional institutions. Readers are asked to accept our assurance that the pronouns 'he' and 'his' imply 'he or she' and 'his or her' throughout.

John Redmond and I are delighted and honoured that Sir Philip Otton has written a Foreword to our book, and that he has been able to endorse it in such generous terms.

John also joins me in expressing our grateful appreciation of the help and encouragement given by Julia Burden of the publishers, and by her colleagues led by Janet Prescott (production) and Sarah-Kate Powell (marketing).

The law is stated from sources available to John and myself at 30 April 1999.

Douglas Stephenson
Salisbury

Addendum: Seventh Edition of the ICE Conditions of Contract

A Seventh Edition of the ICE form is due to be published in September 1999. Although the text of that edition is not available at the time of going to press we are aware of the principal changes that will be made.

The new edition will of course incorporate the many amendments to the Sixth Edition that have appeared since its original publication date of January 1991; and in addition it is expected that the following alterations will be made:

- Under clause 4, the contractor's right to sub-contract will be subject to the engineer's consent, reverting to the form of restriction that applied in the fifth edition.
- Under clause 11, the engineer will be 'deemed to have provided' the contractor with site information instead of being 'deemed to have made available to the contractor' such information.
- Under clause 42(2)(a), the contractor will be entitled to possession of the whole of the site, except where otherwise specified, from the works commencement date, instead of being entitled to 'so much of the site and access thereto as may be required in accordance with the programme'.

The second of these changes represents a further step in clarifying the employer's responsibility for providing site information (see comment on pages 51–2). It should now become a simple and easily determined question of fact whether or not site information was provided to the contractor.

The third change is to be welcomed in that it removes a frequent source of contention to which we refer on pages 83–4: namely the question whether or not the contractor is entitled to occupy parts of the site not immediately required for the construction work itself, but planned to be used for storage of materials etc., until required for the construction works. The contractor will now know that he is entitled to occupy the whole of the site from the commencement date, except for those parts expressly defined as not being so available.

Douglas Stephenson
Salisbury
6 August 1999

CHAPTER ONE
INTRODUCTION

1.1 *Definition of claim*

It is a common perception that claims are a significant feature in all construction projects, whether building or civil engineering. Contractors say that their profit margins are so low that they are obliged to make claims in order to stay in business. Employers try to block claims by introducing ever more onerous amendments to standard forms of contract. A new form of specialist advisor has appeared – 'the claims consultant'.

It may come as surprise to find that the word 'claim' is not defined in either the ICE Conditions Sixth Edition or the ICE Minor Works Conditions. The side-note to ICE clause 52(4) refers to 'Notice of Claims' and the sub-clause lays down the procedure to be followed 'if the contractor intends to claim refixing of rates or any additional payment pursuant to any Clause of these Conditions' other than ordered variations. ICE clause 44 makes reference to the contractor's 'claim' for an extension of time.

The definitions clause of the Minor Works Conditions is also silent on the term. Under it, claims for extension of time are covered by clause 4.4; additional payments are dealt with by clause 6.

The word does not even appear in the Engineering and Construction Contract (originally issued under the name, the New Engineering Contract, and still widely known as that). The innovative term 'compensation event' and the procedures associated with it cover the same ground.

'Claim' is a general term for the assertion of a right to money, property or a remedy. Garner's *Dictionary of Modern Legal Usage* (Oxford University Press, 1995) gives 'to take or demand as one's right' as the primary meaning of the word. This book is concerned with claims for additional money which arise other than under the ordinary contract provisions for payment. It also covers claims for extension of time for completion.

1.2   *Legal basis of claims*

There are four bases on which a claim may be made in law:

- Under the contract conditions themselves.
- For breach of contract, when the contractor's remedy will be damages calculated in accordance with common law principles.
- For breach of a duty arising at common law in tort. This is a general liability, and in principle liability often depends on the defendant having, by act or omission, acted in breach of a legal duty imposed on him by law, so infringing a legal right vested in the claimant and causing him foreseeable damage. Again, the remedy is normally an award of money damages as compensation for the damage done.
- On a quasi-contractual or restitutionary basis, often called a *quantum meruit* claim.

Each of these types of claim will be examined in turn.

1.2.1   Contractual claims

An important feature of both the ICE and Minor Works Conditions is the provision of contractual machinery for dealing with monetary claims under the terms of the contracts themselves. The compensation event procedure is similarly vital to the operation of the New Engineering Contract. These claims arise out of specific provisions of the contract and are dealt with under it by the engineer or project manager. Since they arise under the contract, they are commonly called *contractual claims*.

An example is a claim made under clause 12 of the ICE Conditions which entitles the contractor, in limited circumstances, to claim in respect of delay and extra cost should he encounter certain adverse physical conditions or artificial obstructions as the works progress. The right to payment of any extra cost is dependent on the contractor complying with the notice and related provisions of clause 52(4). This is a general provision which applies to all claims for refixing of rates and 'for any additional payment' under any clause of the contract, other than for ordered variations.

In the Minor Works Conditions contractual claims are provided for in clause 6.1. This refers to the contractor incurring 'additional cost including any cost arising from delay or disruption to the progress of the Works as a result of [one or more of five specified matters]'. These include engineer's instructions requiring the suspension of the works or any part of them. A claim for adverse physical conditions is covered by clause 3.8.

Provisions of this sort are one of the benefits to both parties of using a negotiated standard form contract. The standard form can specify events that give rise to extra payment and provide procedures for settling them. In the examples quoted above there is no blame on the part of the employer or of the engineer – adverse physical conditions are a neutral event and the issue of an instruction suspending the works is merely the exercise of a contractual right. However in both instances the contracts provide for the employer to bear the consequences of delay and extra cost.

Another benefit of these standard form contracts is that both provide procedures for the grant of extensions of time for causes outside the contractor's control, including defaults for which the employer is responsible in law. For instance, the Minor Works Conditions provide for an extension of time for completion for delay caused by the employer's failure to give adequate access to the works or possession of land required to perform the works: clause 4.4(e).

Under the general law, the engineer would have no power or right to grant an extension of time for events outside the control of the contractor or otherwise (*Percy Bilton Ltd* v. *Greater London Council* (1982)). Similarly, the contractor would be under a strict duty to complete on time save to the extent that he is prevented from doing so by the acts or defaults of the employer or those for whom the employer is vicariously responsible. This general rule may be amended by the express terms of the contract and is so amended by clause 44 of the ICE Conditions and clause 4.4 of the Minor Works Conditions, both of which cover neutral events as well as defaults which are the responsibility of the employer in law.

The grant of an extension of time by the engineer under either standard form may or may not also give rise, independently, to a monetary claim under some other clause of the contract. This is emphasised by the wording of clause 6.1 of the Minor Works Conditions with its reference to 'any cost arising from delay *or* disruption'. Both contract forms recognise a distinction between claims for *prolongation* and those for *disruption* – a distinction that is important in practice and which is often misunderstood – and prolongation or disruption claims form part of the recovery of 'cost' under specified clauses in the contract.

1.2.2 Claims for breach of contract

Apart from contractual claims, the contractor may have a claim for damages for breach of contract at common law. This is an entirely different type of claim, the success of which depends upon the contractor proving, on the balance of probabilities, that the employer is in breach of some express or implied term of the contract and that he, the contractor, has suffered loss as a result. In that case, the contractor can recover damages to compensate him, calculated in accordance with common law principles. Claims of this type must in principle be pursued by legal process outside the normal contractual procedures. Such process might be adjudication, arbitration or litigation, with all the inherent uncertainties involved. In fact each of the standard forms considered in this book provides that certain events, that would normally amount to a breach of contract, can nevertheless be dealt with under the contractual machinery as contract claims. This is an attempt to avoid the need for legal disputes.

For example, clause 7(4) of the ICE Conditions entitles the contractor to claim for delay in issuing drawings or instructions by the engineer at the right time. The Minor Works Conditions contain a similar provision in clause 6.1 with its reference to 'delay in receipt by the contractor of necessary instructions drawings or other information'. The engineer's failure to provide information at the right time is a breach of the express provisions of the contract (ICE Conditions, clause 7(1); Minor Works, clause 3.6) for which the employer is vicariously responsible in law: *Neodox Ltd* v. *Swinton and Pendlebury Borough Council* (1958). Both contracts, sensibly, provide a contractual remedy by creating a right for the contractor to have his claim for both delay and extra cost dealt with under the contract by the engineer.

By invoking the contractual mechanism, for example by notifying the engineer of his intention to claim additional payment under clause 52(4) of the ICE Conditions and complying with any other procedural requirements, the contractor secures the right to early payment. His entitlement as decided by the engineer will be included in interim payments or certificates, so creating a right under the contract to a *debt* rather than a claim for damages. Should the employer fail to pay the sum certified as prescribed by the contract, the contractor can suspend his works on site under section 112 of the Housing Grants, Construction and Regeneration Act 1996. He can also proceed rapidly with arbitration in order to obtain an enforceable award or, if an arbitration clause has not been included in the contract, he can sue on the certificate and apply for summary judgment as provided for by Part 24 of the Civil Procedure Rules (previously known as Order 14 of the Rules of the Supreme Court). Additionally, the contractor is given a contractual right to interest on overdue payments: ICE Conditions, clause 60(7); Minor Works Conditions, clause 7.8.

In some cases, therefore, the contractor has a choice as to whether to pursue a claim under the terms of the contract or to claim damages for breach of contract at common law. There appears to be no English civil engineering contract case directly in point – possibly because the civil engineering industry was, until recently at least, rather less litigious than the building industry – but the building contract case of *London Borough of Merton v. Stanley Hugh Leach Ltd* (1985) supports this view. The machinery provided by the contract is not exhaustive of the contractor's rights and remedies.

If the contractual machinery is operated correctly, the contractor will be reimbursed promptly and without the expense and delay of a claim for damages pursued in arbitration or litigation. However, as Mr Justice Vinelott said in *Merton*:

> '[The contractor] may prefer to wait until completion of the work and join the claim for damages for breach of the obligation to provide instructions, drawings and the like in good time with other claims for damages for breach of obligations under the contract.'

There is nothing in the ICE Conditions, the Minor Works Conditions or the New Engineering Contract that excludes such a claim for damages in an appropriate case.

No standard form of contract is a self-sufficient document containing all the contractual obligations of the parties. Each is firmly fixed in the mainstream of the general law.

Both sets of conditions contain their own definition of the 'Contract' and at first sight these definitions suggest that the 'Contract' as defined is completely self-contained. The ICE Conditions, clause 1(1)(e), define the 'Contract' as meaning,

> the Conditions of Contract Specification Drawings Bill of Quantities the Tender the written acceptance thereof and the Contract Agreement (if completed).

5

The corresponding provision in the Minor Works Conditions, clause 1.2, says that the 'Contract'

> means the Agreement if any together with these Conditions of Contract the Appendix and other items listed in the Contract Schedule.

(The omission of any punctuation in these quotations – and in any other quotation from either set of conditions – is that of the draftsmen of the Conditions and not our omission.)

Nevertheless, although these definitions might suggest to non-lawyers that everything they need to know is contained in the 'Contract' as so defined, the printed conditions and other documents referred to contain only the *express terms* agreed by the parties. In some cases the courts will write a term into the contract, usually in order to make it commercially effective. These are called *implied terms,* and breach of an implied term may also give rise to a claim for damages. Clause 49(3) of the Fifth Edition of the ICE Conditions refers explicitly to 'any obligation expressed *or implied* on the Contractor's part under the contract'. Although Clause 49(3) of the Sixth Edition merely refers to 'neglect or failure by the Contractor to comply with any of his obligations under the Contract', the change in wording makes no difference to the principle. Terms can equally be implied on the part of the employer.

The subject is a complex one and is discussed in Chapter 6. At this stage, it is sufficient to state that most claims by contractors for breach of such terms are based on implied terms that the employer (and his agents) will not hinder or prevent the contractor from completing the contract in accordance with its terms. The standard forms (particularly the Minor Works Conditions) do not deal exhaustively with every conceivable act of the employer, the engineer and so on which might hinder or prevent the contractor from completing the contract expeditiously. Terms to cover the situation will therefore be implied.

1.2.3 Claims in tort

The law of tort is that part of the law that imposes a civil duty generally, and breach of that duty may give rise to a claim for damages. The contractor here would be alleging breach of a duty arising at common law other than in contract. In practical civil engineering terms the most important tort is that of negligence. This developed rapidly between 1970 and 1990, although recent cases have tended to a narrowing of negligence liability save where there is actual resultant physical damage. Economic loss is now recoverable only in very limited circumstances, such as claims involving negligent advice, and *Pacific Associates Inc.* v. *Baxter* (1988) (see Chapter 2) establishes that, except in the most unusual circumstances, a contractor cannot pursue a claim for negligence to recover economic loss against the engineer in his role as contract administrator. However, there remains the possibility of claims for misrepresentation, either at common law or under the Misrepresentation Act 1967. Such claims are likewise discussed in Chapter 6.

1.2.4 Quasi-contractual claims

Quasi-contractual or restitutionary claims can arise in a number of ways, and the most common example is a claim on a *quantum meruit* ('as much as it is worth'). This is a claim for the value of services rendered or work performed and may arise:

- under a contract where no price is fixed or where the agreement is to pay a reasonable sum; or
- where there is no valid and enforceable contract between the parties but the law imputes an obligation to pay a reasonable sum for work done or services performed – a claim in restitution.

A restitutionary *quantum meruit* claim may arise, for example, where work is done on the basis of a letter of intent and there is no contractual liability. So in *British Steel Corporation* v. *Cleveland Bridge & Engineering Co Ltd* (1981) the defendants were involved in construction work in Saudi Arabia and needed to be supplied with cast steel nodes, for the supply of which they negotiated with BSC. They sent a letter of intent to BSC in the following terms:

'We are pleased to advise you that it is [our] intention to enter into a sub-contract with your company, for the supply and delivery of the steel castings which form the roof nodes on this project.... We understand that you are already in possession of a complete set of our node detail drawings and we request that you proceed immediately with the works pending the preparation and issuing to you of the official form of sub-contract.'

BSC started work. Some days later the defendants informed BSC of the delivery sequence they required, this being the first time it was stated that the nodes were to be manufactured in a particular order. Despite the problems caused, BSC continued to manufacture the nodes. No 'official form of sub-contract' was ever sent, nor did the parties agree on price or delivery dates. In the event, disputes developed and the defendants failed to pay. BSC claimed £229 832.70 from the defendants as the price of the nodes, and framed their claim alternatively in contract and upon a *quantum meruit*. Mr Justice Robert Goff (as he then was) held that no contract had come into existence between the parties on the basis of the letter of intent and BSC's performance of the work, but BSC were entitled to be paid upon a *quantum meruit*. This in fact worked to BSC's advantage because, since there was no binding contract, there was no basis for the defendants' counterclaim that, in breach of contract, BSC had allegedly delivered the components late and out of sequence.

According to the High Court of Australia in *Pavey & Matthews Pty Ltd* v. *Paul* (1986), in order to establish a claim to a *quantum meruit* in restitution – as opposed to a contractual *quantum meruit* discussed below – the claimant must establish that:

- there is no subsisting valid and enforceable contract between the parties,
- he has performed work without payment, which has conferred a benefit,
- the benefits conferred were not intended to be done gratuitously; and
- the benefit has been accepted (actually or constructively) by the defendant.

It is thought that the position is the same in English law.

A letter of intent may in fact, in some circumstances, give rise to a binding contract: *Turriff Construction Co Ltd* v. *Regalia Knitting Mills Ltd* (1971). It is a question on the facts of each case as to whether a letter of intent gives rise to any and, if so, what liability.

If at all possible, the uncertainty of letters of intent should be avoided. In *British Steel Corporation* v. *Cleveland Bridge & Engineering Co Ltd* (1981) Mr Justice Robert Goff clearly summarised the legal position:

'There can be no hard and fast answer to the question whether a letter of intent will give rise to a binding agreement: everything must depend on the circumstances of the particular case. In most cases, where work is done pursuant to a request contained in a letter of intent, it will not matter whether a contract did or did not come into existence because, if the party who has acted on the request is simply claiming payment, his claim will be based on a *quantum meruit*, and it will make no difference whether that claim is contractual or quasi-contractual... But where, as here, one party is seeking to claim damages for breach of contract, the question of whether any contract came into existence is of crucial importance.'

A quantum meruit claim can also arise where work has been done by a contractor under a contract without there being any express agreement as to the price. This is called a contractual *quantum meruit* and is a claim under the contract to recover reasonable remuneration where there is no agreement as to price or where the parties have agreed that the price shall be 'a reasonable sum'. For example, in *Amantilla Ltd* v. *Telefusion plc* (1987), a builder agreed to do work for a price of £36 626 plus VAT. As work progressed extensive extra works were agreed to be done, the builder saying 'off the cuff' that the extras would 'certainly be in the region of £60 000 or so at the very least'. All the work was completed satisfactorily and interim payments of £53 000 had been made. A receiver and manager of the plaintiff company was appointed and, despite subsequent correspondence and meetings when the defendants stated that they would shortly submit an offer 'somewhere between' £10 000 and £132 000, only a further interim payment of £5000 was made. The defendants offered a further sum of £2000 in full and final settlement, which would have brought the total paid for the original contract work and extras to £60 000. This offer was not accepted. The builder was held entitled to recover upon a *quantum meruit*.

The expression *quantum meruit* is also used to describe a contractual claim for a reasonable sum where, for example, the contract rates have ceased to be applicable or where the pricing provisions are void. In *Constable Hart & Co Ltd* v. *Peter Lind & Co Ltd* (1978) the parties entered into a sub-contract in the FCEC standard form for the surfacing of a by-pass. The sub-contract incorporated the claimants' quotation, which had provided for the price to 'remain fixed price until 3 June 1975; any work carried out after this date to be negotiated'. The sub-contract works were not completed by 3 June 1975, and no agreement was reached on the price for the work carried out thereafter. The Court of Appeal held that the sub-contract was subject to an implied term that the rates for work to be carried out after 3 June 1975 would be reasonable rates, and the matter was remitted to the arbitrator.

A not dissimilar situation arises where the employer has rendered performance of the contractor's obligations impossible, for example by failure to issue a direction or a variation order. This was the situation in *Holland Hannen & Cubitt (Northern) Ltd* v. *Welsh Health Technical Services Organisation* (1981), where the employers were held liable to pay sub-contractors for window assemblies. Judge John Newey QC said:

> '[The architect's] failure to issue a variation instruction in 1974 rendered Cubitts' and Crittalls' obligation to supply windows pursuant to the entire contracts into which they had entered impossible of performance. Crittalls went on to provide windows "at their own risk", meaning that, if the windows should still leak, no payment would be made to them, but in the expectation that, if the windows did not leak, they would be paid ... [The architect and employer] by their words and conduct, agreed to that arrangement. I think that the result was a new contract between Crittalls and [the employer], to which Cubitts were not parties, that, in consideration of Crittalls providing window assemblies for the hospital, [the employer] would pay them a *quantum meruit* ... [The] windows now in the hospital are better in design than those which Crittalls agreed to supply [originally], and they include more, and more expensive, materials. *Prima facie* therefore, I think that Crittalls might expect to recover a larger amount than the sub-contract sum.'

Merely making the contractor's job difficult, as opposed to impossible, does not justify the abandonment of the contract rates and procedures in favour of a *quantum meruit*. In *McAlpine Humberoak Ltd* v *McDermott International Inc* (1992) the plaintiff sub-contractor succeeded in convincing the Official Referee as the court of first instance that the issue of a very substantial number of variations early on in the contract effectively changed the whole nature of the contract, and indeed frustrated it. Instead of being entitled to the contract price, adjusted in accordance with the contract to take account of the numerous variations, the sub-contractor was entitled to a *quantum meruit*. This was overturned in the Court of Appeal in which it was held that

'If we were to uphold the judge's finding on frustration, this would be the first contract to have been frustrated by reason of matters which had not only occurred before the contract was signed, and were not only well known to the parties, but had also been provided for in the contract itself.'

Nevertheless, many contractor's claims are, in fact, made on a *quantum meruit* basis. This is not what the standard forms of contract require, although clause 6.1 of the Minor Works Conditions, for example, provides that:

> if the Contractor carries out additional works or incurs additional cost ... the Engineer shall certify and the Employer shall pay to the Contractor such additional sum as the Engineer after consultation with the Contractor considers *fair and reasonable.*

This is effectively a *quantum meruit* approach, although the engineer will use the contract rates as a guide to what is 'fair and reasonable', and the contract rates may represent in effect a ceiling on the amount payable.

This will not be the case where, for example, a contractor is claiming on a *quantum meruit* where the employer has wrongfully brought the contract to an end: see *Renard Constructions (ME) Pty Ltd v. Minister for Public Works* (1992). Similarly the sub-contract rates have no relevance where a main contractor has wrongfully represented to a sub-contractor that work ordered as a variation constitutes part of the main contract works: see *Costain Civil Engineering and Tarmac Construction Ltd* v. *Zanen Dredging and Contracting Co Ltd* and Chapter 11.2 below.

1.2.5 *Ex gratia* claims

Ex gratia ('out of kindness') claims are those which the employer has no obligation to meet. Contractors often put them forward merely because they are losing money. Sometimes *ex gratia* payments are made to settle or compromise a claim rather than go to the expense of contesting it in litigation or arbitration.

What is effectively a contractual provision for an *ex gratia* payment is to be found in clause 56(1) of the Government General Conditions of Contract for Building and Civil Engineering (GC/Works/1, Edition 3), which gives the Authority (the employer) a *discretionary* power to terminate the contract at any time. This power is not conditional on default by the contractor. If it is exercised, clause 58(4) provides:

> If the contractor is of the opinion that his unavoidable losses or expense directly due to determination have not been fully reimbursed by sums paid or agreed then he shall refer the circumstances to the Authority, who shall make such allowance, if any, as in his opinion is reasonable.

In effect, all this sub-clause appears to do is to enable the employer to make *ex gratia* payments if he so decides. The purpose of the provision is not at all clear, except that the contractor's financial entitlement under earlier sub-clauses is very limited indeed.

CHAPTER TWO
CLAIMS AND THE ENGINEER

2.1 Engineer's role and functions in claims settlement

The engineer is not a party to the construction contract. He is usually under a separate contract, formal or informal, with the employer, to advise upon and to control all technical aspects of the project from inception to completion. He is also the 'machinery' to supervise the contractor's performance of the contract and to decide what payments shall be made. This necessarily involves valuing the work done, and deciding on variations, additions and omissions as the work proceeds. Hence the engineer must form a view on the proper interpretation of the contract provisions, certifying interim payments and the final account.

Clause 1(1)(c) of the ICE Conditions defines who the engineer is: he is the 'person, firm or company appointed by the Employer to act as Engineer for the purposes of the Contract'. Clause 2.1 of the Minor Works Conditions provides for the appointment of a 'named individual to act as Engineer'.

The engineer is the agent of the employer for these purposes. He is bound in contract to his principal, the employer. The terms and scope of his agency are those to be inferred from what he is employed to do under the terms of the contract. As Judge John Davies QC put it at first instance in *Pacific Associates Inc* v. *Halcrow International Partnership* (1987):

> 'Short of any fraud or dishonesty on his part, which would be manifestly outside the scope of his agency, his principal will be liable for the acts of his agent, subject to any right of recourse he may have against his agent for indemnity.'

An underlying assumption behind both sets of conditions is that the engineer must bring an independent professional judgment to bear when deciding certain questions and must act fairly and impartially as between the employer and the contractor. Many clauses of both the ICE and Minor Works Conditions require the engineer to exercise his discretion on matters which affect the contractor's interests.

The major example is certificates. Clause 60(2) of the ICE Conditions says that 'within 28 days of the date of delivery to the Engineer ... of the Contractor's monthly statement the Engineer *shall certify* and the Employer *shall pay*' the amount certified to the contractor. Clause 7.3 of the Minor Works Conditions is to the same effect. It is quite clear that the employer must not interfere with the certifying process; for example, by instructing the engineer not to issue a certificate until some dispute is resolved.

This is illustrated by *Hickman & Co* v. *Roberts* (1913), where an employer instructed his architect not to issue a certificate until he had received the contractor's account for extras. The House of Lords held that the employer had improperly interfered with the exercise of the architect's powers as certifier and the contractor was entitled to recover even in the absence of a certificate. The Australian decision of *Perini Corporation* v. *Commonwealth of Australia* (1969) is a more recent example of this principle. There the Supreme Court of New South Wales held that it was an implied term of a construction contract that the employer would do all that he reasonably could to ensure that the engineer properly performed his duties as certifier. Certainly the courts will protect the contractor's rights, even in the absence of a certificate: *Croudace Ltd* v. *London Borough of Lambeth* (1986).

The principle, therefore, is that in these areas the engineer must act fairly, taking into account not only the interests of his client, the employer, but also, to an equal extent, those of the contractor.

Although many of these matters are of a financial nature – agreeing or determining rates for varied work or additional payment in respect of delays, etc. for which the employer is responsible – questions may also arise where the engineer may have to decide whether or not the standard of workmanship and material is in accordance with the specification. The employer is usually not present at the time when the engineer is called upon to deal with questions of this nature. Indeed, in many major civil engineering projects the employer is a corporate body to which the engineer is responsible as agent, no other individual being appointed to represent the employer's interests.

It follows that in dealing with claims and other contentious matters the engineer must perform his two separate and often conflicting roles concurrently. Inevitably, this dichotomy of roles can lead to suspicion on the contractor's part that the engineer may be lacking impartiality in allowing himself to be influenced excessively by his own persuasive powers in his role as 'employer's agent'. The suspicion is enhanced where the subject-matter of the dispute has a bearing upon the engineer's own work; where, for example, a claim is based upon alleged delay in the issue of the engineer's drawings.

It is not surprising that under these contract procedures disputes arise and have sometimes to be referred to arbitration. What is much more surprising is that procedures apparently so fundamentally in conflict with the rules of natural justice lead to so few disputed decisions. It should be noted particularly that the engineer has only limited powers of delegation under the ICE Conditions, clause 2(4), and many important questions cannot be delegated at all. Clause 2.4 of the Minor Works Conditions appears to be less restrictive, although even here the final decision is that of the named engineer should the contractor be dissatisfied with the delegate's decision.

It is of the greatest importance that, when dealing with claims before or after they become the subject of disputes, the engineer should act, and be seen to act, impartially. Even where a claims submission is poorly presented or is incomplete, an out-of-hand rejection does nothing to enhance the engineer's reputation for impartiality in the eyes of the contractor or indeed of any reasonable person. A detailed explanation of the inadequacies of the presentation, and providing an opportunity to remedy them, may encourage the contractor to believe that the engineer's objective is not solely to reject his claims, but to do so only when a careful and impartial examination of the facts shows that course to be justified.

Where the contractor so requests, we suggest that the engineer should allow him an opportunity to make oral representations in support of his claim; even when, after careful study of a written submission, the engineer has formed an opinion that a claim is invalid. This is especially important where the contractor is unable to command the resources and skills of the larger companies, and is reluctant to seek independent outside advice.

It is not unusual for a contractor to complain that the engineer is being dominated by the employer. This is a situation that the engineer should not allow to develop. It is the engineer who is required to reach a decision on contentious matters between the employer and the contractor. Although the engineer may and should take note of the employer's contentions (which he may himself have to express), he should not allow the influence of his client, the employer, to override his own impartial judgment where differences arise.

We have heard of cases where employers have allegedly instructed engineers not to issue variation orders, grant extensions of time or settle monetary claims. A blanket instruction of this type is undoubtedly a breach of contract. In the building contract case of *Croudace Ltd* v. *London Borough of Lambeth* (1986) a not dissimilar situation arose, since outside quantity surveyors replied to a contractor's claim that they had been instructed by Lambeth that all claims for additional payment must be approved by the employer to whom they had sent the claim submission 'for their guidance in this matter'. Lord Justice Balcombe, in the Court of Appeal, said that the outside quantity surveyors

> 'were, of course, acting as agent for the Chief Quantity Surveyor of Lambeth, who had certain independent duties to perform under the contract ... Unless this letter is intended to mean that on this matter Lambeth's Chief Quantity Surveyor was not prepared to delegate his functions ... it is not clear to me what right Lambeth had to issue this instruction.'

An employer who wished to have the power to issue instructions could amend the Conditions accordingly, but it is not desirable that this should be done. Local authority standing orders may require that variations, etc. above a certain level should be sanctioned by the appropriate committee; if this is the case, the standing orders should be incorporated into the contract.

Clause 51(1) of the ICE Conditions distinguishes those variations that are *necessary* for the completion of the works, from those that are *desirable* for the completion and/or improved functioning of the works. The engineer has a *duty* to order variations falling within the first category, and is *empowered* (but not *required*) to order those falling within the second category. An increase in the depth of a foundation necessary where an unsuitable stratum is encountered at the designed depth would clearly constitute a *necessary* variation; while an increase in the floor area of a warehouse in order simply to increase its capacity is an example of a *desirable* variation. In the latter example there can of course be no objection to the employer deciding whether or not to sanction the variation when its cost is known.

Express provision is made in the ICE Sixth Edition, clause 2(1)(b), but not in its predecessor, for limiting the engineer's authority 'specified in or necessarily to be implied from the contract'. Particulars of any such limitation must be set out in the appendix to the form of tender, and any requisite approval is deemed to have been given where the engineer exercises any such authority.

Although this provision has been criticised on the ground that it may in some circumstances fetter the engineer in the exercise of what should be his impartial judgment, others contend that some degree of control over the engineer by the employer is usual, and that the new provision merely ensures that the contractor is aware of the degree of control exercised.

The important point is that under both the ICE and Minor Works Conditions the named engineer has independent duties to perform. In the public sector, the role of internal auditors is sometimes questioned if the contract requires that the engineer should not issue certificates, etc. until the figures have been checked. Unless the Conditions are amended so as to clarify the role of the internal auditors, this could be said to amount to wrongful interference with the engineer's duties. In the case of certificates, any checking process must be completed within the timetable set down in either set of Conditions.

It should also be understood that there is nothing in the wording of the ICE or Minor Works Conditions to suggest that the position is any different whether the engineer is an employee of the employer, for instance in the public sector, or an independent outside consultant.

2.2 Engineer's liability to the contractor

When dealing with claims for extensions of time or for additional cost and in issuing certificates and so forth, the engineer's duty under a civil engineering contract, and indeed under any other form of construction contract, is to act fairly. That duty gave rise, in the early 1970s, to a notion that when performing such functions the engineer could be described as a 'quasi arbitrator' and as such was immune from suit. The notion was firmly rejected by the House of Lords in *Sutcliffe* v. *Thackrah* (1974), in which it was held that an architect who negligently over-certified payments under a building contract was liable to the employer in negligence. In that case, the employer had paid against the negligently over-valued certificates and the contractor became insolvent. The same principle clearly applies to an engineer acting under the ICE or Minor Works Conditions.

The more interesting question is whether the engineer owes a duty of care to the contractor in respect of under-certification or in the exercise of his other functions under the contract requiring him to act fairly. The position now appears to be that the contractor cannot successfully bring a claim in tort against the engineer for negligent under-certification that results in purely economic loss. This follows from the decision of the Court of Appeal in *Pacific Associates Inc* v. *Baxter* (1988), where the contract incorporated the FIDIC International Civil Engineering Conditions and the contractor brought an action against the engineer claiming damages for economic loss allegedly caused by negligent certification and failure to act fairly and impartially in administering the contract. (See also Chapter 6.4 below.)

The Court of Appeal, upholding the trial judge, struck out the claim as disclosing no cause of action. Their lordships held that there is no simple and unqualified answer to the question: 'Does the engineer owe a duty to the contractor in tort to exercise reasonable care and skill?' The Court was of the opinion that the question can only be answered in the context of the factual matrix, especially the contractual structure against which such a duty could be said to arise: and, having examined the previous case law, concluded that such a duty could not be imposed on the engineer.

The contract in fact contained a special clause (PC 86) which provided that 'neither the engineer nor any of his staff shall be in any way personally liable for the acts or obligations under the contract'. However this does not appear to have been a decisive factor in the decision, although the fact that there was an arbitration clause in the contract was, in the Court's view, an important factor. Lord Justice Purchas expressed the view that, even if the disclaimer had not been included, 'the provisions of [the arbitration clause] would be effective to exclude the creation of any direct duty upon the engineer towards the contractor'.

In summary it seems at present that, under civil engineering contracts, the contractor is unlikely to be able to establish a claim in tort against the engineer in order to recover purely economic loss – the more so as *Pacific Associates* has been followed in other common law jurisdictions – although it is possible (if unlikely) that such a claim might lie if there were no arbitration clause in the contract. If the engineer unfairly promotes the employer's interests by low certification or merely fails properly to exercise reasonable care and skill in his certification, the contractor has the right against the employer to have the certificate reviewed in arbitration; or, following upon the decision of the House of Lords in *Beaufort Developments (NI) Ltd* v. *Gilbert-Ash NI Ltd and Others* (1998), in litigation.

There appears to be nothing in the wording of the ICE Conditions or the Minor Works Conditions to suggest that the position and liabilities of the engineer are any different from those of an architect under a building contract in respect of matters in which the contract requires him to act impartially, such as the issue of certificates, the settlement of claims for additional cost and the grant of extensions of time. In *Davy Offshore Ltd* v. *Emerald Field Contracting Ltd* (1991), Judge Thayne Forbes QC stated:

> '[It] is clear that the obligation to act fairly is concerned with those duties of the architect/engineer which require him to use his professional judgment in holding the balance between his client and the contractor. Such duties are those where [he] is obliged to make a decision or form an opinion which affects the rights of the parties to the contract, for example, valuations of work, ascertaining direct loss and expense, granting extensions of time, etc. When making such decisions pursuant to his duties under the contract, the architect/engineer is obliged to act fairly.'

The ICE and Minor Works Conditions proceed on the assumption that a clear distinction is to be drawn between the acts of the engineer as contract administrator/agent of the employer and his acts when acting as impartial certifier holding the balance between the employer and the contractor.

However, it is now clear that, as a result of the decision of the Court of Appeal in *Pacific Associates*, under both the ICE and Minor Works Conditions the contractor has no direct right of action against an unfair or negligent engineer in tort.

2.3   Delegation of the engineer's powers

The ICE Conditions provide, in clause 1(1)(c), for the engineer to be a named individual, firm or company. Under clause 2(2) however, where the engineer so defined is not a single named chartered engineer, the contractor must be notified within seven days of the award of the contract of the name of such a chartered engineer who will act on behalf of the engineer. This is subject, of course, to the power of the employer to replace him by another should the need arise. Clause 2.1 of the Minor Works Conditions is to the same effect. In general, it is to be hoped and indeed expected that such a replacement would not be made except where it becomes necessary to do so – for instance, where the named engineer dies or becomes seriously ill. This is because the engineer's personal reputation for fairness and impartiality is a factor taken into account by many contractors in pricing their tenders.

Where the employer is a public authority or a corporation having its own engineering department, it is customary for the chief engineer of that department to be named as the engineer under the contract. In many cases, however, that named engineer, being at the head of a large organisation, finds it necessary to delegate his powers almost in their entirety to a member of his staff, usually termed the project engineer. In this way, the named engineer may in practice have little or no involvement in the administration and control of the project, remaining aloof from the day-to-day activities and taking an active part only where serious problems or disputes arise.

It is debatable whether or not such an arrangement is in the best interests of both of the parties to the contract. It has the merit of enabling the named engineer to fulfil more naturally the role of impartial engineer than where he has been involved personally in control and administration, thereby leaving his project engineer free to act as employer's agent without the inhibitions that would otherwise arise. However, there may be many occasions during the period of construction when the project engineer is required to make impartial decisions, knowing that the named engineer expects him to use the authority he has been given in all but the most serious of differences or construction problems.

Where the engineer appointed by the employer is a partnership of consulting engineers, one named partner should be appointed to be the engineer under the contract. It is sometimes argued that such an appointment implies a greater degree of independence, and therefore of impartiality, than where the employer appoints his own in-house employee to be engineer under the contract.

Conversely, it may be argued, with some force, that the reverse is true. The chief engineer to a public authority or corporation may know that the tenure of his appointment is less likely to be affected by any disenchantment his employers may have with a decision in favour of the contractor than where the engineer is a consultant, whose appointment may be terminated on completion of the contract in question or, indeed, in extreme circumstances, before that completion. However such action would undoubtedly support a submission that the contractor had been treated unfairly.

Delegation of the engineer's authority to a project engineer is usually informal in the sense that the project engineer acts for and on behalf of the engineer, and signs his letters and instructions as project engineer. Where such an arrangement is made, the contractor retains a right of access to the engineer in person under clause 2(7) of the ICE Conditions.

Some public authorities issue letters of delegation stating that the named engineer will exercise certain powers himself and that he has delegated other powers to other named persons. This seems to follow the pattern of the ICE Conditions, clause 2(4), and is far better practice than having the named engineer merely sign, on a routine basis, documents prepared by his delegates. As we discuss below, a great many very important questions cannot be delegated at all under the ICE Conditions. If the chief engineer of a public body, for example, is routinely appointed as the named engineer (whether in his personal name or by description of his office), all those non-delegable questions must be referred to him as the named engineer for his personal decision.

Under the Minor Works Conditions, there is no similar restriction because of the wording of clause 2.2, which says that 'the Engineer may delegate ... *any* of the powers of the Engineer herein provided that prior notice in writing is given to the Contractor'. Even here, however, as a matter of good practice the named engineer should consider personally any objection that the contractor might make to his delegate's decisions.

Clause 2(1) introduces into the Sixth Edition a definition that was, under the Fifth and earlier editions, conspicuous by its absence: namely, a definition of the duties and authority of the engineer. He *shall* carry out the duties and *may* exercise the authority specified in or necessarily to be inferred from the contract. (Limitation of his authority by the employer is referred to in section 2.1 above.)

Clause 2(3) defines the functions and powers of the engineer's representative, more commonly known as the resident engineer or RE. The sub-clause makes it clear that the functions of the RE, unless specifically extended, are limited to watching and supervising the construction and completion of the works. In the Fifth Edition the RE also had to supervise 'maintenance': a term that has now been superseded by the more accurate description of the period following completion of the works as being the 'defects correction' period. Although the Sixth Edition does not expressly so provide, it is to be presumed that the RE must also exercise his supervisory function during the defects correction period. The RE's powers, as defined, are very limited, and it is rare for him to perform only those limited functions. Provision is made in clause 2(4), for the engineer to delegate some of his functions to the RE, either generally in respect of the contract or specifically under certain defined clauses. Where such authority is given to the RE, the contractor must be notified in writing.

However, the engineer's power to delegate his functions is circumscribed by the proviso to clause 2(4)(c). The engineer has no power to delegate his authority under:

- Clause 12(6)   To decide that physical conditions or artificial obstructions could not have been foreseen by an experienced contractor
- Clause 44   To determine extensions of the time for completion
- Clause 46(3)   To request accelerated completion of the works
- Clause 48   To issue a certificate of completion
- Clause 60(4)   To issue a final certificate
- Clause 61   To issue a defects correction certificate
- Clause 63   To certify abandonment of the contract, etc.
- Clause 66   To decide upon matters in dispute

The inclusion in the above list of clause 12(6) but not clause 12(5) indicates that the RE is empowered to reject a claim under clause 12 but not to allow such a claim. However the distinction is of minor importance because a contractor whose clause 12 claim is rejected by the RE has recourse to the engineer under clause 2(7).

Within the above limits, it is essential that the engineer should implement the provisions of clause 2 by deciding upon and notifying the contractor of the appointment of an RE and of the extent of his delegated powers. In practice however the engineer, while intending that the RE should have at least limited authority to issue minor variation orders and instructions and to agree varied rates, sometimes fails to notify the contractor of the extent of those powers. It is suggested that in such circumstances the contractor should, preferably at the first site meeting held prior to the commencement of construction, request the engineer to define the extent of the delegated powers to be exercised by the RE and his assistants, pursuant to clauses 2(4) and 2(5).

Failing such formal delegation of the engineer's functions to the RE, the contractor must realise that a variation or instruction issued by the RE (other than an instruction to comply with the require-ments of the contract) is *ultra vires*. The powers and duties of the engineer are exercisable only by the named engineer or under the delegation powers of clause 2(4). Action upon such an instruction, and payment for any varied work ordered, should be refused on the ground that it was not properly authorised. Upon receipt of such an instruction, the contractor should point out to the RE that it requires to be validated by the engineer himself. Such action may lead to a proper definition of the extent of the RE's delegated powers.

Under the Minor Works Conditions, the position is less complex. Clause 2.1 provides for the employer to appoint a named individual to act as engineer. Clause 2.2 contains the powers of delegation:

> The Engineer may appoint a named Resident Engineer and/or other suitably experienced person to watch and inspect the Works and the Engineer may delegate to such person in writing any of the powers of the Engineer herein provided that prior notice in writing is given to the Contractor.

That delegation may be general or specific, but even here it is suggested that the named engineer might consider reserving to himself the more important of the powers conferred upon him by the contract. Indeed, the Guidance Note makes the situation plain: 'in all normal circumstances the Engineer would not be expected to delegate his powers' as to

- adverse physical conditions (clause 3.8)
- extensions of time (clause 4.4)
- certifying completion (clause 5.4) or
- the final certificate (clause 7.6).

Clause 2(6)(c) introduces into the Sixth Edition of the ICE Con-ditions a provision first incorporated in the Minor Works Condi-tions under which the engineer or the RE is required, on the contractor's request, to specify in writing the authority for any instruction given. Clause 2(7) entitles the contractor to appeal to the named engineer against any delegate's instruction. A similar pro-vision appeared in clause 2(4) of the Fifth Edition, while in the

Minor Works Conditions the corresponding provisions appear in clause 2.4. The wording of clause 2(7) of the Sixth Edition is important:

> If the Contractor is *dissatisfied by reason of any act or instruction of the Engineer's Representative or any other person responsible to the Engineer* he shall be entitled to refer the matter to the Engineer for his decision.

This would appear to be a condition precedent to service of a notice of dispute under clause 11.2 of the Minor Works Conditions, which says that no notice of dispute

> may be served unless the party wishing to do so has first taken any step or invoked any procedure available elsewhere in the Contract in connection with the subject matter of such dispute and the other party or the Engineer as the case may be has:
> (a) taken such step as may be required, or
> (b) been allowed a reasonable time to take any such action.

2.4   Engineer's decision under ICE Conditions – clause 66(3)

2.4.1   Procedure

The procedure for the settlement of disputes arising from the ICE Conditions is defined in clause 66. Sub-clause (2) provides a definition of the time at which a dispute is deemed to have arisen, namely when one party serves on the engineer a *Notice of Dispute*. This the party may only do after having invoked any procedures available elsewhere in the contract in connection with the subject matter of the dispute, and after having allowed a reasonable time for the engineer to take the actions provided for in those procedures. Sub-clause (3) provides for settlement of the dispute by the engineer, who must state his *decision* in writing. This must be done within one month if the dispute arises before the issue of a completion certificate, or within three months if substantial completion of the whole of the works has been certified. That decision, unless referred to conciliation or arbitration by either party within the strict time limits provided for in sub-clauses 66(5) and 66(6), becomes final and binding upon the parties.

In general, a contractor wishing to pursue claims is well advised to ensure that those claims are well prepared and presented; that any additional information sought by the engineer is provided, and that every effort is made, by written or by oral persuasion, to reach a fair settlement by negotiation. Only when such efforts have proved to be of no avail, and only if the contractor, having heard the engineer's reasons for his rejection, remains convinced of the validity of his claims, should action be taken under clause 66.

Many contractors regard a request for an engineer's decision under clause 66(3) as committing them to arbitration and closing the door to further negotiation. However, while action under the clause should not be taken lightly or frivolously, such action does not, of itself, result in any financial or other commitment on the part of the contractor. It certainly does not signal the end of attempts to negotiate a settlement: on the contrary, it may sometimes have the opposite effect, by indicating to the engineer the seriousness of the position and thereby promoting a more determined effort to negotiate a settlement.

2.4.2 Need for unquestionable impartiality

It should be recognised that on occasions some engineers do not appear to be as impartial as they ought to be. For example, where the cost of a project is already in excess of the engineer's estimate or the original budget cost (for whatever reason), and where the employer has legitimately urged the engineer to accept his, the employer's, contentions regarding the matters in dispute, the distinction between hearing the employer's case and being dominated by the employer is impossible to define. In such circumstances, the engineer may have serious doubts as to his own ability to be impartial and may welcome the reference of the dispute to a person whose impartiality cannot be questioned, i.e. a conciliator or arbitrator.

Similarly, some engineers who know that they face criticism from their clients for allowing costs to exceed earlier estimates may believe that an arbitrator's decision to allow a claim will be more palatable to the employer than would be the engineer's own decision to the same effect. This is in our view a wrong approach to the problem.

The engineer's decision is not final and may be challenged in arbitration which, like marriage, is not a step to be taken lightly or inadvisedly. The ICE Conciliation Procedure, which was first published in 1988 and was amended in 1997, was initially incorporated in the Minor Works Conditions and later in the Sixth Edition. It is a move towards a structured attempt to resolve disputes by negotiation rather than arbitration.

2.4.3 Criticisms of clause 66 procedure

The first stage under clause 66, namely that of referring the dispute to the engineer for his decision, has not surprisingly been criticised by some commentators. The dispute itself arose from the engineer's decision: why, it is therefore argued, should the engineer (again asked to make that decision but this time under clause 66) alter the ruling he has already given?

An argument sometimes advanced is that the request for a decision under clause 66 concentrates the engineer's mind upon the seriousness of the position and may result in his reviewing the facts and revising his earlier decision. That contention ignores the duty of the engineer to concentrate his mind when the matter is first presented to him, in order that he may come to a fair decision. He could not have properly performed his duty to act impartially if it is only when faced with reference under clause 66 that he gives proper consideration to the matters in question.

A further criticism of the first stage of the clause 66 procedure is that it causes delay. Where the engineer takes the full period he is allowed to respond to a request for his decision, the applicant suffers a delay of three months (or, in the case of a request during the period of construction, one month) before he is able to invoke the arbitration agreement.

A third and more real difficulty sometimes arises where an engineer contends that his rejection of a claim, given more than three months earlier, constituted a decision under clause 66 which the contractor, by failing to require that the matter be referred to arbitration within three months of receipt of the decision, is deemed to have accepted as being final and binding. Hence, it is argued, the contractor is out of time and has no means of reopening the matters in question.

2.4.4 Validity of clause 66 decisions

This situation was considered by the Court of Appeal in *Monmouthshire County Council* v. *Costelloe & Kemple Ltd* (1965), which arose under the Fourth Edition of the ICE Conditions. In that case, it was held that the initial rejection of the contractor's claims was not a decision under clause 66 because until the time of that rejection there was no 'dispute or difference' upon which a decision could be given.

The engineer must make it quite clear where he intends that a statement shall be a decision under clause 66. The engineer's letter in *Monmouthshire County Council* merely rejected the claims and did not decide them. The engineer had written to the contractor enclosing his 'observations and comments' on them. With one exception, he rejected all the claims saying, 'I cannot agree with or consider the claim', and gave his reasons. There was thus no 'decision' under clause 66(1). The case established a number of points, some of which remain applicable to the Sixth Edition.

- There must first be a 'dispute or difference', which supposes a clear rejection of the claim in quite unequivocal words: or, where it is the employer who gives notice of dispute, a clear acceptance by the engineer of the claim.
- Once a dispute has arisen, it must be referred to the engineer 'who shall state his decision in writing', notifying both employer and contractor.
- The timetable in sub-clause 66(6) must be observed; if the engineer fails to respond, then the matter must be referred to arbitration within the prescribed period if the applicant wishes to pursue his claim. The same situation arises where the engineer has given a valid clause 66 decision which is not acceptable to one party or the other.
- If the engineer gives his decision within the prescribed period, then the disgruntled party has three calendar months from *receipt* of the engineer's decision in which to refer the matter to arbitration.
- Clause 68 – dealing with service of notices – applies to the engineer's decision under clause 66. It is best, in the case of notices served by post, to ensure that recorded delivery or registered post with advice of delivery is used, though this is not required by the clause. The curious legal presumption that a letter posted by first-class delivery is deemed to have been served on the next working day following posting certainly cannot be relied upon in this day and age. It is thought that clause 68 contemplates that the notice will only be effective when it is actually delivered.

2.4.5   Sixth Edition changes

Although the dispute resolution procedure remains substantially similar to that of the Fifth Edition a number of minor improvements have been introduced, some of which were foreshadowed in the Minor Works Conditions. For example the introduction, in clause 66(2), of a definition of the time at which the dispute is deemed to arise is often of considerable value in saving costly, time-consuming, and unnecessary contention. In particular, the definition will be relevant where a sub-contractor seeks to invoke the arbitration procedure defined in clause 18(7) of the CECA Form of Sub-Contract, while the main contractor seeks to rely on the provision of clause 18(10)(d) of the CECA Form, under which the main contractor may require the dispute to be dealt with jointly with a dispute under the main contract. Before that improvement was introduced a main contractor could reject a sub-contractor's notice of arbitration on the ground that the dispute fell to be dealt with under the dispute provisions of the main contract, and thereafter fail to proceed with the main contract arbitration. (But see also Chapter 11, section 11.8.) Again in common with the Minor Works Conditions, provision has been introduced for conciliation under the ICE Conciliation Procedure (1988). This provision is optional; it may be invoked by either party after the engineer has given his clause 66 decision, or after the time within which he should do so has elapsed.

Whilst the definition of when a dispute arises is effective in the context of the ICE Conciliation Procedure and the arbitration agreement, there is real doubt as to its efficacy in relation to adjudication because of the requirements of the Housing Grants, Construction and Regeneration Act 1996. This is discussed in Chapter 7 below.

2.5   *The position of the engineer under the Minor Works Conditions*

The Minor Works Conditions make provision for disputes settlement under clause 11, and this clause (rather more sensibly) contemplates a different procedure. Once again, there must be a 'dispute or difference' between the contracting parties – 'including a dispute as to any act or omission of the Engineer'. Clause 11.2 then goes on to provide that a written notice of dispute cannot be served on the other party unless the disputant 'has first taken any step or invoked any procedure available elsewhere in the Contract in connection with the subject matter of such dispute' and the engineer or the other party has failed to take the required steps or has failed to act 'within a reasonable time'. In the situation of a claim being made this does not appear to mean that a formal decision of the engineer is required in the sense that it is under the ICE Conditions.

For example, clause 3.8 (dealing with delay and extra cost caused by adverse physical conditions) envisages that, in the absence of agreement between the contractor and the engineer, the engineer is to 'determine the fair and reasonable sum to be paid' to the contractor for any additional cost. If the contractor alleges that the sum is not 'fair and reasonable', his remedy is to serve notice of dispute under clause 11.2; and so it is with additional payments under clause 6 and, indeed, with extensions of time under clause 4.4.

There is no requirement that the engineer should give a further formal 'decision' on the dispute or difference, and the optional provision for conciliation (under clause 11.3) means that claims disputes should be settled with the minimum of delay. However, it is to be noted that paragraph 13 of the Guidance Notes published with the ICE on the Minor Works Conditions makes some very important points relating to disputes:

(1) The option provided by the Conciliation Procedure under Clause 11.3 is intended to provide a means whereby disputes can be settled with a minimum of delay by obtaining an independent recommendation as to how the matter in dispute should be settled. The conciliation is complete when the conciliator has delivered his recommendations and, if any, his opinion to the parties.

(2) It is normally expected that the party serving a Notice of Dispute under Clause 11.2 will at the same time serve notice in writing either under Clause 11.3 (Conciliation) or Clause 11.5 (Arbitration).

(3) It should be noted that if within the prescribed period of 28 days after service of a Notice of Dispute neither party has made a request in writing for the dispute to be referred to a conciliator nor has served a written Notice to Refer requiring the dispute to be referred to arbitration then the Notice of Dispute becomes void. It is then open to either party to continue the dispute by serving a fresh Notice of Dispute unless the first Notice of Dispute was served within the 28 days allowed under Clause 7.7 (Final certificate) in which case the final certificate becomes final and binding and no further dispute in respect of the Contract is possible.

2.6   *Reference to arbitration – the engineer's role*

For many years, it was the intention of the ICE Conditions that arbitration should be deferred until completion or alleged completion of the works. Clearly, this objective is based upon the sensible view that it is better to deal with all possible disputes that may arise from a contract in a single arbitration, thereby reducing the time and the cost that would otherwise be incurred in referring all of the disputes separately. Furthermore an arbitration conducted before completion of the works is likely to cause inconvenience and disruption to the works, because many of the staff of both engineer and contractor might need to be involved in preparing the claims and the defence, and/or be required as witnesses of fact at the hearing.

However, the danger of a rigid adherence to this policy is that where a contractor incurs additional costs (allegedly through some cause for which the employer is responsible) and is unable to obtain prompt payment, he might be unable to complete the contract works, and hence be unable to comply with the condition precedent to his invoking arbitration. To allow for such a possibility, the original version of the Fifth Edition, published in 1973, provided two grounds upon which the applicant was entitled to 'immediate' arbitration:

- firstly, in the case of a dispute arising under clause 12 of the Contract; and
- secondly, where the dispute arose from the 'withholding by the Engineer of any certificate'.

The meaning of those words was considered by the Court of Appeal in *A E Farr Ltd* v. *Ministry of Transport* (1960), where it was held that the failure of the engineer to certify payment of more than £15 000 of a £20 000 claim constituted 'withholding a certificate' and therefore entitled the contractor to immediate arbitration of the dispute.

The ICE drafting committee appears to have taken the view that the *Farr* decision prevented the effective imposition of any bar to immediate arbitration. Accordingly, clause 66(5)(c) of the revised version of clause 66 in the Fifth Edition (published on 10 June 1985) and clause 66(6)(a) of the Sixth Edition provide expressly for arbitration to proceed notwithstanding that the works are not complete or allegedly complete.

It is, nevertheless, to be hoped that parties will not unnecessarily invoke arbitration before completion of the works, since to do so must incur the risk that there may have to be more than one arbitration arising from the contract. In circumstances where a party alleges that the other party, by invoking arbitration at an early stage, has incurred unnecessarily the increased costs of multiple references, it would be open to that party to draw the arbitrator's attention to its allegation and to invite him to use his discretion in his award of the unnecessary part of the costs. The option of adjudication in contracts entered into since 1 May 1998 should further discourage premature arbitration.

It should be recognized that the 'dispute or difference' referred to arbitration arises from the contract between the employer and the contractor. The engineer, although having a function defined within that contract, is not a party to it and in consequence is not directly involved in the arbitration.

The engineer is, however, one of the key witnesses in that he has a detailed knowledge of both design and construction of the project. Furthermore, in a number of cases the dispute may arise from an act or omission of the engineer himself which the contractor alleges constitutes improper exercise of his discretionary powers under one or more of the many clauses of the contract under which he is given such powers. Where the engineer's actions are called into question in this way, he will clearly have a key role in defending those actions. The RE and members of his staff are likely to be called as witnesses of fact. The engineer and his staff are also likely to be involved in a similar way in any adjudication procedures.

Since 1973, what is now clause 66(9) of the Sixth Edition (clause 66(6) of the Fifth Edition) clarifies the role of the engineer where a dispute is referred to arbitration. The engineer may be called to give evidence before the arbitrator, notwithstanding that he has given a decision on the dispute under clause 66(3). The engineer is indeed an essential component of the machinery for disputes settlement under clause 66.

The engineer's role is the same under the Minor Works Conditions where the disputes procedure is invoked.

2.7 *Power to revise engineer's decisions*

The engineer is empowered under many clauses of the ICE Conditions to make decisions that affect not only the execution of the works but also the payments due to the contractor. The fact that the engineer is appointed and paid by the employer must sometimes give grounds for questioning his impartiality; especially in dealing with matters that arise from his own performance. For example, a claim in respect of delay to the works caused by the late issue of drawings must imply deficiencies within the engineer's own organisation which might influence him towards rejection of the contractor's claim.

In order to give the contractor a right of recourse to a person whose impartiality cannot be questioned the engineer's decisions are expressly subject to review by an arbitrator appointed under clause 66 of the ICE conditions. Sub-clause 66(8) includes the provision that such arbitrator 'shall have full power to open up review and revise any decision opinion instruction direction certificate or valuation of the Engineer'. It was thought that the prime purpose of the provision was to enable decisions of fact (including the engineer's opinions) to be reviewed by a person suitably qualified in engineering, and only such a person. For example a decision whether or not, in a claim arising under clause 12, the subsoil conditions encountered might reasonably have been foreseen by an experienced contractor usually requires a sound knowledge of civil engineering, based upon academic qualification combined with substantial experience as engineer or as contractor. Decisions on questions of law, such as the true construction of terms of the contract, are already open to review by the arbitrator; hence no express power is needed.

In *Northern Regional Health Authority* v. *Derek Crouch Construction* (1984) the Court of Appeal held that under a corresponding provision in the JCT form of building contract, the power given expressly to the arbitrator to 'open up, review and revise' decisions of the architect, was available exclusively to the arbitrator and not to the court. The logic underlying that decision is not hard to find: the arbitrator is impliedly a skilled construction practitioner, while the judge is not.

In *Balfour Beatty Civil Engineering* v. *Docklands Light Railway* (1996) the contract between the parties was based on the ICE Conditions, Fifth Edition, but with two important alterations. The engineer was replaced by 'the employer's representative', and clause 66, which includes the arbitration agreement, was deleted. The Court of Appeal held that it did not have power to 'open up, review and revise' decisions of the employer; the court's powers being 'limited to special circumstances established in law as justifying interference by the courts'. It followed that the decisions of the employer's representative, whose impartiality was, having regard to his title, even more questionable than that of an impliedly impartial 'engineer', could not be revised. Sir Thomas Bingham MR answered a question that had been submitted as a preliminary issue in his own words: 'The court may grant appropriate relief to the Contractor if and to the extent that it proves breaches of contract by the employer'.

What had become known as the *Crouch* principle, namely that the power given expressly under the terms of the ICE and JCT forms of contract to an arbitrator to review and revise decisions of the engineer or architect is not available to the court, was overturned by the House of Lords in *Beaufort Developments (NI) Ltd* v. *Gilbert-Ash NI Ltd and Others* (1998), when *Crouch* was expressly overruled and *Balfour Beatty* expressly disapproved. The basis of the ruling was that decisions of the architect under the terms of the JCT form are not, in general, expressed as being conclusive. Similar reasoning would of course apply to decisions of the engineer under the ICE form.

The parties are however free to agree on the machinery for establishing their obligations, and the court could not substitute different machinery. In *Beaufort Developments* their lordships quoted an example:

> 'The contract may say that the value of the property or the question of whether the goods comply with the description shall be determined by a named person as an expert. In such a case, the agreement is to sell at what the expert considers to be the value or to buy goods which the expert considers to be in accordance with the description. The court's view on these questions is irrelevant.'

If, as appears to be the case, the intention of the words used in clause 66(8)(a) of the ICE Conditions is that decisions of the engineer on technical matters (i.e. matters of fact, including expert opinion, as distinct from matters of law) should be subject to review only by a person having suitable knowledge of engineering, then it is impliedly open to the three bodies responsible for drafting the ICE conditions, namely the ICE, the ACE, the CECA and the FCEC (since succeeded by the CECA) to amend the conditions so as to implement that intention. In the light of *Beaufort Developments* this could be done by introducing a provision that the engineer's decisions are final unless and until revised by an arbitrator appointed pursuant to clause 66(8). Where disputes are referred to litigation rather than arbitration such a provision would not inhibit the court from dealing with questions of law: but it would be necessary for a plaintiff to prove breach of contract in order for the court to over-turn the engineer's decision.

CHAPTER THREE
CONTRACTUAL CLAIMS

3.1   The meaning of 'cost'

The ICE Conditions contain a definition of 'cost' for the purposes of claims arising under the contract itself. The definition in clause 1(5) of the Fifth Edition is not particularly helpful:

> The word 'cost' when used in the Conditions of Contract shall be deemed to include overhead costs whether on or off the Site except where the contrary is expressly stated.

The Sixth Edition gives a more explicit definition:

> The word 'cost' when used in the Conditions of Contract means all expenditure properly incurred or to be incurred whether on or off the Site including overhead finance and other charges properly allocatable thereto but does not include any allowance for profit.

Unfortunately, the word 'cost' is not used exclusively or con-sistently throughout the Conditions. In some clauses 'cost' is qualified by an adjective, and in other provisions undefined words of similar – but possibly not identical – meaning are used.

In clauses 7(4), 12(3), 13(3), 14(8), 17, 20(3), 21(1), 31(2), 36(2), 36(3), 38(2) and 39(2) of the Sixth Edition Conditions the word 'cost' is used on its own, while in clause 42(3) there are references to 'additional cost'; 'extra cost' appears in clause 40(1). Clause 12 provides for payment to the contractor of the 'costs which may reasonably have been incurred by the Contractor, by reason of such conditions or obstructions together with a reasonable percentage thereto in respect of profit'.

In contrast, clauses 20(3), 32 and 59(4) refer to work at the 'expense' of the employer. Clause 52(1) uses the term 'value' in respect of varied work which impliedly includes additional work. This word is also used in clauses 49(3), 51(3), 55(2) and 59(2) and, by implication, in clause 18.

In contrast to earlier editions, in which a percentage addition to the cost of work performed was allowable under clause 12, but under no other clause, the Sixth Edition provides for the addition of profit to all costs incurred in a clause 12 claim, and in many other clauses, including clauses 13, 14, 31(2) and 40(1). Except where profit is expressly included in a clause giving such an entitlement, the definition of 'cost' invalidates any claim for profit.

The position is different where the word 'expense' is used. It is submitted that, on the true construction of the contract, where 'expense' is used to describe payments borne by the employer, it must include profit as well. There are many examples from case law, and in the context 'expense' is to be distinguished from 'cost' and is at least arguably wider in its meaning: *Chandris* v. *Union of India* (1956). It is also thought that 'expense' may well include interest not earned if the contractor uses his own capital as opposed to operating on an overdraft basis. The *Concise Oxford English Dictionary* definition of 'expense' is 'cost incurred; payment of money'.

In the Minor Works Conditions the definition of 'cost' is similar to that in the Sixth Edition – again expressly excluding profit – the most significant difference being the inclusion of 'other charges (including loss of interest) properly allocatable thereto'. Again, as in the Sixth Edition, clause 3.8 (adverse physical conditions and artificial obstructions) expressly includes 'cost ... together with a reasonable percentage addition in respect of profit'. It can only be assumed that the reasoning underlying this special case is that additional work carried out in a clause 12 or clause 3.8 situation would have been priced at rates including their own profit element had that work been foreseen and included in the bill of quantities.

'Value' is used in relation to clause 52, in which rules for evaluation are given. Since the basic principle of clause 52 is the application or adjustment of contract rates to varied work, and since such rates should generally include an element of profit, it appears that the principle used is that payment for additional work should include a profit element. On the other hand, losses suffered through delays for which the employer is responsible are reimbursed under the contract at cost only, unless express provision is made for profit, even though a claim for loss of profit would form a head of a claim for damages for breach of contract at common law.

3.1.1 Site overheads

Site overheads cover all of those site costs that are not directly related to particular items of work. Generally, they include:

- Salaries and wages of all 'non-productive' staff, i.e. the agent, site engineers and quantity surveyors, secretaries, clerks, the general foreman, walking gangers, plant fitters, storemen, chainmen, tea boys and cleaners.
- Office erection, hire, removal and other costs (including heating, lighting, sanitation, cleaning, telephone rental and call charges, stationery and postage).
- Security, watching, lighting and traffic control.
- Provision of stores, workshops, canteens, latrines, etc. (including erection, hire, removal, heating, lighting and cleaning costs).
- Construction, maintenance and removal of site access roads.
- Haulage of plant to and from the site.
- Contract insurances and performance bond.
- Provision and maintenance of small tools.
- Subsistence and travelling costs and allowances.

The majority of the above items are of a time-related nature, but there are a few, such as office erection and haulage of plant to and from the site, in which the whole or part of the cost is a lump sum. The latter must obviously be excluded when calculating the weekly site oncost, which cost may in general be claimed in respect of delays for which the employer is responsible.

3.1.2 Off-site overheads

Off-site overheads include the salaries and wages of head-office and, where applicable, regional management, estimating, surveying, administrative and accounting staffs, together with costs associated with their employment – office accommodation, facilities and travelling.

As a generality, it can be said that there is – or should be – a fairly constant relationship between annual turnover and annual head-office overheads, so that the latter may be expressed as a percentage of on-site costs, which percentage remains roughly static from year to year. Hence the estimator knows the percentage to be added to his total of site costs in order to provide an estimate of total cost, to which only profit – usually also calculated as a percentage of total site costs – needs be added in order to arrive at the amount of the tender. However that 'fairly constant' relationship has been upset from time to time, for example during the exceptionally deep and prolonged recession in the early to mid 1990s when some contractors were unable to adjust rapidly to falling turnover. It is to be expected that disputes may have arisen, and may continue to arise, from this cause.

Where the contractor is engaged solely in contracting, that activity is the only source from which off-site overheads may be recovered. The head-office contribution is usually added into the estimated site cost as a percentage addition, and its recovery is on a time-related basis. If the contract is prolonged, there is a shortfall of contribution; and where the prolongation results from some cause for which the employer is financially responsible the contractor should be able to recover the additional prolongation costs incurred.

It was not until the Fifth edition of the ICE Conditions was published in 1973 that the definition of 'cost' expressly included off-site overheads. Earlier editions included no such provision; but the contractor's entitlement to additional off-site overheads as part of his costs had been recognised at common law. In the Canadian case of *Shore & Horwitz Construction Co Ltd* v. *Franki of Canada Ltd* (1964) the Supreme Court of Canada held that in circumstances of prolongation the contractor was entitled to recover off-site overheads as a head of claim because the contractor was prevented from earning revenue to defray his overheads during the period of delay.

3.1.3 The formula approach

In 1970 Ian Duncan Wallace QC, the author of the Tenth Edition of *Hudson's Building and Engineering Contracts*, expressed in that edition the opinion that 'a contractor's loss [of head-office overheads and profit] from an extended contract period must be a proportionate extension of [the HO/profit] percentage of the contract sum and the loss calculated in this way is a real loss (provided the true percentage used can be determined)'. Surprisingly perhaps in a lawyer having a meticulous regard for the precise wording of a contract, Duncan Wallace did not separate head-office costs from profit. His reasoning was that 'some contractors do consciously apply a breakdown of the percentage as between head-office and profit, but for the purpose of assessing the loss due to delay in completion, the division is not theoretically important'. Hence he derived what has become known as the 'Hudson formula' for calculating the additional off-site overheads/profit during a prolongation period as being:

$$\frac{\text{HO/Profit percentage}}{100} \times \frac{\text{Contract sum}}{\text{Contract period (weeks)}} \times \text{Period of delay (weeks)}$$

The author of *Hudson* entered two caveats to the use of the formula:

- firstly that the contractor did not habitually underestimate his costs when pricing, so as to inflate his profit percentage; and
- secondly that there had been no change in the market, so that work of at least the same level of profitability would have been available to him at the end of the contract period.

The first of these two caveats may where necessary be dealt with by the adoption of an alternative formula set out in *Emden's Building Contracts and Practice*, 8th edn, vol. 2, p.N/46, in which the prolongation costs are quoted as being:

$$\frac{h}{100} \times \frac{c}{cp} \times pd$$

where

h = head-office percentage arrived at by dividing the total overhead cost and profit of the contractor's organisation as a whole by total turnover

c = the contract sum

cp = contract period in weeks

pd = period of delay in weeks.

The Emden formula differs from the Hudson formula only in that it takes the HO/oncost percentage from the contractor's overall organisation, instead of the percentage built into the tender. It may in some cases be more realistic in its approach than its rival – on the basis that it uses figures actually achieved by the contractor's organisation, in preference to estimated profitability which may have been unduly optimistic. On the other hand it is sometimes at least arguable that the percentage included in the tender for oncost and profit reflects special circumstances of the particular contract, such as unusual risks or unsocial timing, and is more correctly applicable than the average percentage profit and oncost calculated from the contractor's annual accounts. The authors are, for example, aware of a case where a contractor whose usual profit margin was of the order of 1% of total site costs, adjusted slightly in either direction to take account of market conditions, issued a successful tender which included a profit margin of 40% – albeit that it was for a relatively small contract. The reason for the contractor's departure from his usual practice was that the main element of the work, which involved the construction of a pedestrian underpass beneath a railway line, was to be carried out during a 72-hour possession of the line, commencing shortly before midnight on 24 December.

Both the Hudson and the Emden formulae contain a mathematical error in that the HO/oncost percentage is applied to the contract sum, which sum does of course already include the result of adding that percentage. Hence the formulae should be corrected by substituting 'contract sum minus HO/oncost addition' for 'contract sum'.

Secondly the formulae, for the reason referred to above, provide for enhancement of the profit element, to which the contractor is not always entitled. In *J Crosby & Sons Ltd* v. *Portland Urban District Council* (1967), which arose under an earlier edition of the ICE Conditions, it was emphasised by Mr Justice Donaldson (as he then was) that no award should be made containing a profit element unless this is expressly permitted by the contract. Although the entitlement to profit as an element of claims arising from certain contractual provisions has been extended to cover many more clauses in the ICE Conditions giving rise to claims, it remains necessary to ensure that profit is not enhanced in respect of prolongation unless the clause from which the claim arises expressly so provides.

Case law since the development of the formula approach to the calculation of off-site overheads during a prolongation period may be summarised as follows.

In the Canadian case of *Ellis-Don Ltd* v. *The Parking Authority of Toronto* (1978) the contractor claimed that 3.87% of the contract price had been included for off-site overheads and profit. In allowing a claim based on the Hudson formula for prolongation costs the judge made specific findings in relation to the assumptions mentioned based on the evidence led. Hence the case is not authority for the view that the formula is of blanket application; it must be backed up by evidence. An excellent critique of the formula is contained in the editorial commentary to the *Building Law Reports* report of *Ellis-Don* as follows:

'The circumstances of each contract and of each contractor should ... be considered carefully before a decision is made whether or not the Hudson formula is applicable ... [W]hether the claim is presented as damages for breach of contract or upon a contractual right to recover loss ... the person liable for the claim, if established, or those acting on his behalf will wish to see all the documents relevant to the checking of the claim such as the make-up of the tender, and the internal head office and job records and accounts relating to overheads and to actual and anticipated profits and profit levels both in and on either side of the period of delay.'

> *Building Law Reports*, vol. 28, pp102–3.

In *Tate & Lyle Food & Distribution Co Ltd* v. *Greater London Council* (1982) the plaintiffs had claimed 2.5% of the total damages for managerial time spent on remedying an actionable wrong. Mr Justice Forbes accepted that such a claim was in principle recoverable as a head of damages, but the claim failed because the plaintiffs had failed to prove their loss by proper records, and no acceptable yardstick or evidence for assessing the amount claimed was given. Mr Justice Forbes said:

'While I am satisfied that this head of damage can properly be claimed, I am not prepared to advance into an area of pure speculation when it comes to quantum.'

The Hudson formula has been discussed in a number of Australian cases. For example, in *State of South Australia* v. *Fricker Carrington Holdings Ltd* (1987), there was direct discussion of whether the Hudson formula could be used in the absence of evidence that its use was appropriate. The Appeal Court referred to the trial judge's view that 'there is no doubt that this is a formula very frequently used in cases of this type ... I have always understood the formula in *Hudson* to be a good starting-off point in this type of case'. The Appeal Court continued:

'I am sure that parties in dispute frequently use the formula. Often they will agree to its use. But if they do not agree to its use I think that evidence must be called to prove that its use is appropriate. It is not a formula in a statute or in a regulation. One cannot take a formula said by a textbook writer to be usually used and assert that arbitrators must use it. If this matter is reconsidered consistently with my opinion the arbitrator should not use the formula without proof that it is appropriate.'

The position is the same under English law and was accurately expressed by Sir William Stabb QC in *J F Finnegan Ltd* v. *Sheffield City Council* (1988):

'It is generally accepted that, in principle, a contractor who is delayed in completing a contract, due to the default of his employer, may properly have a claim for head-office or off-site overheads during the period of delay, on the basis that the workforce, but for the delay, might have had the opportunity of being employed on another contract which would have had the effect of funding overheads during the overrun period.'

In that case a formula approach was adopted by the court but, although the judgment gives the impression that the judge was adopting the Hudson formula, he actually applied the Emden formula.

More recently, in *St Modwen Developments Ltd* v. *Bowmer & Kirkland Ltd* (1996), Judge Fox-Andrews QC, on the basis that the formula approach was acceptable at a time when the construction market was in a buoyant state, dismissed an application by the employer to set aside that part of an arbitrator's award that was, by agreement of the parties' experts on quantum, evaluated by the Emden formula.

Conversely, in *City Axis Ltd* v. *Daniel P Jackson* (1998), Judge Toulmin QC dismissed a claim by the plaintiff for off-site costs based on the Emden formula. The judge stated:

'There was no direct evidence that City Axis turned away other work or of specific projects which City Axis might have been able to bid for successfully but for having resources tied up in this contract ... The only evidence relating to a specific project is that they were able to obtain one in Bruton Street in September 1995 very soon after the work on Mr Jackson's project had been completed.'

With respect, the judge in *City Axis* appears to have ignored the normal process by which a contractor obtains work; namely by the submission of a tender that is in due course accepted by the employer thereby creating the contract. Hence the opportunity to 'turn away other work' does not arise. In order to obtain the evidence referred to in the judgment it would have been necessary for the contractor to spend time and money tendering for work that he has no intention of performing; and, having obtained evidence in the form of an acceptance that he was able to obtain work, then to extricate himself from the contract so created.

The author of *Hudson* wisely based his criterion in respect of the contractor's ability to obtain work on there having been 'no change in the market': and that principle was adopted by the judge in *St Modwen Developments*.

To summarise the present state of the law, the authors are of the opinion that in principle the two best known formula approaches, namely Hudson or Emden, are valid in appropriate circumstances, but that where their use is challenged, evidence to substantiate their validity is necessary. Such evidence may not be readily available because it is necessarily hypothetical – relating as it does to what would have happened had the contractor not been delayed on the project on which the claim arises. Furthermore the contractor cannot reasonably be expected to seek work, without having any intention of performing it, simply in order to demonstrate that he could have obtained work had he not been otherwise engaged. A contractor who did so would not endear himself to potential employers; and indeed might find himself in breach of a contract created by the acceptance of a tender submitted for the sole purpose of demonstrating his ability to obtain work.

Having regard to these difficulties it seems to the authors that a properly prepared claim based on a Hudson, Emden, or a similar formula, and supported by evidence of appropriate percentage allowances for off-site overheads and for profit ought to succeed if the employer is unable to refute that evidence. The percentages allowable must be considered separately under the two categories, namely overheads and profit.

Overheads may typically be of the order of 3–5%, although values outside that range may be valid in special circumstances – for example the use of some novel or sophisticated form of construction involving an abnormal degree of risk or managerial supervision. Again, it may be said that one element of off-site overheads, namely estimating costs, is inappropriate in a prolongation claim, where no estimating is required. On the other hand it could be argued that a contractor who is unable to take on new work because of delay on an existing contract cannot lay off his estimating staff without incurring some penalty, such as redundancy costs – quite apart from the likely difficulty in recruiting staff when conditions so warrant.

Where the overhead percentage included in the tender is challenged by the employer the contractor should be prepared to adduce evidence as to its validity, for example by reference to his annual accounts and to any special circumstances that distinguish the contract from his more usual activities.

The profit element is allowable only where the contract so provides: under the ICE Conditions, Sixth Edition, the ambit of clauses giving an express entitlement to profit as an addition to cost has been widened to include 12(6), 13(3), 14(8), 31(2), 40(1), 42(3). Certain other clauses imply an entitlement to profit, for example claims arising from variations under clauses 51 and 52. Whether the profit percentage, where allowable, should be that built into the tender (as in *Hudson*) or that shown in the contractor's accounts (as in *Emden*) must depend on circumstances. Profit percentages vary widely with market conditions and other considerations. It is open to the contractor to argue that an unusually high percentage was built into the tender to allow for special circumstances or conditions, to verify that percentage by producing the estimator's pricing notes; and hence to argue that the percentage shown in his annual accounts is inapplicable. Similarly it may sometimes be arguable that the converse of Ian Duncan Wallace's caveat relating to market conditions applies – that the tender provided an unusually low profit percentage because of the scarcity of work at the time since when market conditions have improved so as to justify an enhanced percentage on current work.

All of the above considerations indicate that the formula approach is at best a very approximate method of assessment of overheads and profit. Nevertheless it is submitted that in the light of the express entitlement to overheads and, in some cases, profit as an element to be added to cost, the tribunal responsible for determining entitlement should be prepared to examine the evidence adduced by both parties and to base its decision on that evidence. Dismissal of a claim on the ground that it is speculative is, it is submitted, not a valid option. The formula used must of course be adjusted to correct the mathematical error referred to above.

3.2 The basis of contractual claims

Work carried out as described in, or as reasonably to be inferred from, the contract documents provides no basis for a claim. The contractor is deemed in law to be experienced and hence to be able to foresee, at the time of making the contract, what the average experienced contractor could foresee as being likely to be required in the performance of that contract. Thus it follows that in order to establish a contractual claim the contractor must be able to show that the work he was required to do, or the conditions under which he was required to do it, differed in some way from what he expected, or should have expected, at the time of making the contract.

The changes that may arise fall in general into one of two main categories:

- variations, and
- delays for which the contractor is not himself responsible.

The categories are closely related in that variations can and frequently do cause delays: equally a delay caused, for example, by non-availability of a construction material could result in the issue of a variation obviating the need for that material.

Both origins of claims are dealt with in detail in later chapters. It is, however, important at this stage to distinguish between delays for which the employer is responsible and those for which the contractor is wholly or partially responsible.

The contractor must usually, as part of his responsibilities, provide the labour, plant, materials, temporary works (including the design of those works) and supervision necessary for purposes of construction, and in accordance with the specification and other requirements of the contract. Delays caused by deficiencies in any of these items are in general the contractor's responsibilities, except where he can show some other root cause – for example a late variation that requires the provision of materials or plant not readily available.

The employer is responsible for making the site available to the contractor and for the issue, through his engineer, of drawings, documents, decisions, variations, instructions, approvals, consents, and payment and other certificates required by the contract.

These are considered in detail in the following sections.

3.3 Documents mutually explanatory

Clause 5 of the ICE Conditions states:

The several documents forming the Contract are to be taken as mutually explanatory of one another and in case of ambiguities or discrepancies the same shall be explained and adjusted by the Engineer who shall thereupon issue to the Contractor appropriate instructions in writing which shall be regarded as instructions issued in accordance with Clause 13.

Clause 5 deals with the not infrequent occurrence of ambiguities or discrepancies in the documents forming the contract. In this context 'documents' clearly includes drawings issued at the time of making the contract, together with specifications, bills of quantities, and any other written information made available at the time of tendering. The engineer has a duty to explain and adjust the ambiguity or discrepancy, and to issue appropriate instructions.

It is open to the contractor to claim that his pricing was based upon the information overruled by the engineer in giving such instruction, and that in consequence he has incurred additional cost in having to do something that was not envisaged at the time of tendering. Where such a situation arises, the instruction is equivalent to an instruction under clause 13. (See section 3.6 below.)

Under the Minor Works form, clause 2.3(g) empowers the engineer to give instructions for 'the elucidation or explanation of any matter to enable the Contractor to meet his obligations under the Contract', but this does not give rise to a contractual claim for additional cost.

3.4 Late issue of further drawings and instructions

3.4.1 ICE Conditions

Clause 7 of the Sixth Edition states:

(1) The Engineer shall from time to time during the progress of the Works supply to the Contractor such modified or further Drawings Specifications and instructions as shall in the Engineer's opinion be necessary for the purpose of the proper and adequate construction and completion of the Works and the Contractor shall carry out and be bound by the same.

If such Drawings Specifications or instructions require any variation to any part of the works the same shall be deemed to have been issued pursuant to Clause 51.

(2) Where sub-clause (6) of this Clause applies the Engineer may

require the Contractor to supply such further documents as shall in the Engineer's opinion be necessary for the purpose of the proper and adequate construction completion and maintenance of the Works and when approved by the Engineer the Contractor shall carry out and be bound by the same.

(3)   The Contractor shall give adequate notice in writing to the Engineer of any further Drawing or Specification that the Contractor may require for the construction and completion of the Works or otherwise under the Contract.

(4)   (a)   If by reason of any failure or inability of the Engineer to issue at a time reasonable in all the circumstances Drawings Specifications or instructions requested by the Contractor and considered necessary by the Engineer in accordance with sub-clause (1) of this Clause the Contractor suffers delay or incurs cost then the Engineer shall take such delay into account in determining any extension of time to which the Contractor is entitled under Clause 44 and the Contractor shall subject to Clause 52(4) be paid in accordance with Clause 60 the amount of such cost as may be reasonable.

(b)   If the failure of the Engineer to issue any Drawing Specification or instruction is caused in whole or in part by the failure of the Contractor after due notice in writing to submit drawings specifications or other documents which he is required to submit under the Contract the Engineer shall take into account such failure by the Contractor in taking any action under sub-clause (4)(a) of this Clause.

(5)   One copy of the Drawings and Specification furnished to the Contractor as aforesaid and of all Drawings Specifications and other documents required to be provided by the Contractor under sub-clause (6) of this Clause shall at all reasonable times be available on the Site for inspection and use by the Engineer and the Engineer's Representative and by any other person authorized by the Engineer in writing.

(6)   Where the Contract expressly provides that part of the Permanent Works shall be designed by the Contractor he shall submit to the Engineer for approval

(a)   such drawings specifications calculations and other information as shall be necessary to satisfy the Engineer as to the suitability and adequacy of the design and

(b)   operation and maintenance manuals together with as completed drawings of that part of the Permanent Works in sufficient detail to enable the Employer to operate maintain dismantle reassemble and adjust the Permanent Works incorporating that design. No certificate under Clause 48 covering any part of the Permanent Works designed by the Contractor shall be issued until manuals and drawings in such detail have been submitted to and approved by the Engineer.

(7)   Approval by the Engineer in accordance with sub-clause (6) of this Clause shall not relieve the Contractor of any of his responsibilities under the Contract. The Engineer shall be responsible for the integration and co-ordination of the Contractor's design with the rest of the Works.

Clause 7(1)

Clause 7 (1) not only confers a power upon the engineer to issue further drawings and instructions: it imposes upon him a *duty* to do so where such modified or further drawings or instructions are *necessary* for construction and completion of the works.

Clause 7(2)

In the Sixth Edition, clause 7(2) makes corresponding provision for the supply of further documents by the contractor where he is responsible for the design of any part of the permanent works. However, the question of whether or not such documents are *necessary* remains a matter for the opinion of the engineer; although that opinion is subject to review by the arbitrator.

Clause 7(3)

Clause 7(3) imposes an obligation upon the contractor to give adequate notice in writing of any further drawing or specification that the contractor may require for the execution of the works. The third of these sub-clauses is included in order to obviate a difficulty that might otherwise arise where doubt exists as to whether or not additional details are necessary. In some types of construction – especially where the work includes building as distinct from civil engineering – the contractor is expected to use his own expertise in deciding upon matters of detail, while in civil engineering work it is more usual for the engineer to provide full details of his require-ments. Without sub-clause (3), a situation could arise where the contractor incurs delay to his constructional work while awaiting details which the engineer has no intention of providing, or even knowledge that such details are awaited.

Clause 7(4)

Clause 7(4) provides the basis upon which the contractor may claim recompense for delay suffered or cost incurred through failure of the engineer to issue drawings or instructions at a time 'reasonable in all the circumstances'. The meaning of those words was considered by the late Mr Justice Diplock (as he then was) in *Neodox* v. *Swinton & Pendlebury Borough Council* (1958), where it was held that the circumstances to be taken into account included the point of view of the engineer and his staff – implying that the engineer was entitled to reasonable time in which to prepare any modified or further drawings that might be necessary. The contract was made in April 1947, 16 months after the publication of the first edition of the ICE conditions. It was not in a standard form but used the words quoted above, which words have since been used in subsequent editions of the ICE Conditions.

Although the judgment has not been reversed by a higher court, it has been criticised by many commentators on the ground that it misinterprets the intention of the contract. While it is clearly the intention that drawings may be issued after the contract has been let, such drawings should be issued in sufficient time to avoid any delay to the contractor in his planning and construction of the works. Where such delay is incurred, the contractor should be entitled to a corresponding extension of the time for completion and to reimbursement of additional costs incurred.

However there is another aspect of the *Neodox* case that may indicate that the judge had sound legal – if not engineering – grounds for his decision. The claimant, Neodox, had planned to complete the contract work in less than the time provided for in the contract; and the judge decided that Neodox was not entitled unilaterally to impose upon the engineer an obligation to complete his drawings at such time as would enable Neodox to achieve the accelerated programme. Had Neodox sought and obtained, either by qualifying its tender or by negotiation, a reduction in the contractual time for completion, it would have had the right to *require* suitably early completion of the engineer's drawings. That right would however have been accompanied by an *obligation* to complete by the earlier date, failure to do so giving the employer the right to impose liquidated damages from the agreed (contractual) completion date. Hence the notion of 'consideration', seen by lawyers as constituting an essential element of a contract, would have been preserved.

Authority for the view that further drawings etc. should always be issued to the contractor in sufficient time to avoid delay is found in *Holland Hannen & Cubitt (Northern) Ltd* v. *Welsh Health Technical Services Organisation* (1981), where the contract, which was based on the JCT form, was held to include implied terms that the employer and his architect would do all things necessary on their part to enable the contractor to carry out and complete the works expeditiously, economically and in accordance with the contract, and that neither would hinder the contractor in his pursuit of those objectives. However the inclusion in that judgment of the words 'in accordance with the contract' emphasises the need for a contractor to negotiate and agree an earlier completion date than that specified in the contract if he wishes to complete the works in less than the time provided by the contract, and to rely upon the engineer's co-operation in doing so.

Sub-clause (4) also provides for variations resulting from such further drawings and instructions to be evaluated under clause 52. (See Chapter 4, section 4.2.)

Clause 7(6)

In the Sixth Edition, sub-clause (6) makes provision for the supply by the contractor of drawings, etc., both for the purpose of satisfying the engineer as to the adequacy of the design and as a permanent 'as built' record of the work done.

3.4.2   Minor Works Conditions

Under this form the corresponding powers of the engineer to vary the works are contained in clause 2.3. The contractor's right to claim in respect of the late issue of necessary instructions, drawings or other information from the engineer is contained in clause 4.4(a) in respect of extension of the time for completion and in clause 6.1 in respect of additional cost resulting from delay or disruption.

3.5   Unforeseeable conditions and obstructions

3.5.1   Clause 11 – assumptions

Clause 11 of the ICE Conditions Sixth Edition states:

(1)   The Employer shall be deemed to have made available to the Contractor before the submission of his Tender all information on
  (a)   the nature of the ground and subsoil including hydrological conditions and

(b)   pipes and cables in on or over the ground
obtained by or on behalf of the Employer from investigations undertaken relevant to the Works.

The Contractor shall be responsible for the interpretation of all such
information for the purposes of constructing the Works and for any
design which is the Contractor's responsibility under the Contract.

(2)   The Contractor shall be deemed to have inspected and examined
the Site and its surroundings and information available in connection
therewith and to have satisfied himself so far as is practicable and reasonable before submitting his Tender as to

(a)   the form and nature thereof including the ground and sub-soil
(b)   the extent and nature of work and materials necessary for
constructing and completing the Works and
(c)   the means of communication with and access to the Site and
the accommodation he may require

and in general to have obtained for himself all necessary information as
to risks contingencies and all other circumstances which may influence
or affect his Tender.

(3)   The Contractor shall be deemed to have

(a)   based his Tender on the information made available by the
Employer and on his own inspection and examination all as
aforementioned and
(b)   satisfied himself before submitting his Tender as to the correctness and sufficiency of the rates and prices stated by him in the
Bill of Quantities which shall (unless otherwise provided in the
Contract) cover all his obligations under the Contract.

A further step has been taken in the Sixth Edition towards removing
the potential conflict between clauses 11 and 12. Clause 11 is not in
itself a clause from which contractual claims arise, but it has, more
particularly in its earlier wording, provided one of the most common bases upon which such claims are rejected.

The Fourth Edition required the contractor 'to inspect and
examine the site and its surroundings and [to] satisfy himself before
submitting his tender as to the nature of the ground and subsoil (so
far as is practicable) ... and in general [to] obtain all necessary
information (subject as above-mentioned) as to risks contingencies
and other circumstances which may influence or affect his tender'.
In the Fifth Edition the first phrase in parentheses was expanded to
'(so far as is practicable and having taken into account any information in connection therewith which may have been provided by
or on behalf of the Employer)'. Thus even under the softened
wording of the Fifth Edition it remained open to the employer or to
his engineer to base a rejection of a clause 12 claim on a contention

that under clause 11 the contractor should have obtained additional information as to the nature of the subsoil, which would have rendered the conditions encountered foreseeable.

The Sixth Edition makes the contractor responsible only for the *interpretation* of information provided by the employer, and for *inspection and examination* of the site and its surroundings: thereby removing any remaining doubt as to whether or not individual contractors ought themselves to carry out additional site investigation. Responsibility for site investigation now rests firmly on the employer, as indeed it should, and it is to be hoped and expected that thorough site investigations will be carried out by employers before tenders are invited.

The benefits to the employer to be derived from such investigations are several.

- Such a survey provides data enabling the engineer to design his foundations accurately and economically.
- Tenders are likely to be lower where tenderers are able to price the works accurately, knowing the conditions and obstructions they are likely to encounter.
- The cost and delay incurred in dealing with unforeseen obstructions is reduced or eliminated where conditions are known and taken into account in the design of the works.

Additionally the likelihood of valid claims arising under clause 12 is reduced where conditions and obstructions have been discovered and information has been made available to tenderers. For example, a thorough site investigation might indicate that ground conditions are such as to require piled foundations, provision for which is then included in the contract. Tenderers who are aware of that need are able to price the piling – often by a specialist sub-contractor – and to incorporate the work in their overall programme. On the other hand, where the engineer, unaware of the likely need for piles, provides in the contract for spread footings, delay is caused by the need for a subsoil investigation when footings are found to be impracticable, by the need to order a variation to provide for piling, by the need for the contractor to obtain piling plant (or possibly to obtain sub-contract prices for the varied work), and for the work to be executed. In addition to the substantial costs likely to be incurred as a result of the delay, the cost of the varied work itself will inevitably be higher than it would have been had it formed part of a competitive tender.

The engineer should however recognise, when arranging for and collating a site investigation report for inclusion in the tender documents, that his client's interests are not best served where matters of opinion are included in the published report. Many specialists in subsoil investigation provide, in addition to factual information as to borehole logs and test results, their own opinions as to suitable types of foundation and bearing pressures upon them. Advice on such matters should be excluded from the information provided to tenderers, leaving interpretation of the factual information to the tenderer, as envisaged by clause 11.

3.5.2 Clause 12 – claims

Clause 12 of the Sixth Edition provides:

(1)   If during the execution of the Works the Contractor shall encounter physical conditions (other than weather conditions or conditions due to weather conditions) or artificial obstructions which conditions or obstructions could not in his opinion reasonably have been foreseen by an experienced contractor the Contractor shall as early as practicable give written notice thereof to the Engineer.

(2)   If in addition the contractor intends to make any claim for additional payment or extension of time arising from such condition or obstruction he shall at the same time or as soon thereafter as may be reasonable inform the Engineer in writing pursuant to Clause 52(4) and/or Clause 44(1) as may be appropriate specifying the condition or obstruction to which the claim relates.

(3)   When giving notification in accordance with sub-clauses (1) and (2) of this Clause or as soon as practicable thereafter the Contractor shall give details of any anticipated effects of the condition or obstruction the measures he has taken is taking or is proposing to take their estimated cost and the extent of the anticipated delay in or interference with the execution of the Works.

(4)   Following receipt of any notification under sub-clauses (1) or (2) or receipt of details in accordance with sub-clause (3) of this Clause the Engineer may if he thinks fit inter alia

(a)   require the Contractor to investigate and report upon the practicality cost and timing of alternative measures which may be available

(b)   give written consent to measures notified under sub-clause (3) of this Clause with or without modification

(c)   give written instructions as to how the physical conditions or artificial obstructions are to be dealt with

(d)   order a suspension under Clause 40 or a variation under Clause 51.

(5)   If the Engineer shall decide that the physical conditions or artificial obstructions could in whole or in part have been reasonably foreseen by an experienced contractor he shall so inform the Contractor in writing as soon as he shall have reached that decision but the value of any variation previously ordered by him pursuant to sub-clause (4)(d) of this Clause shall be ascertained in accordance with Clause 52 and included in the Contract Price.

(6)   Where an extension of time or additional payment is claimed pursuant to sub-clause (2) of this Clause the Engineer shall if in his opinion such conditions or obstructions could not reasonably have been foreseen by an experienced contractor determine the amount of any costs which may reasonably have been incurred by the Contractor by reason of such conditions or obstructions together with a reasonable percentage addition thereto in respect of profit and any extension of time to which the Contractor may be entitled and shall notify the Contractor accordingly with a copy to the Employer. The Contractor shall subject to Clause 52(4) be paid in accordance with Clause 60 the amount so determined.

The principle underlying clause 12, namely that the cost of dealing with unforeseeable physical conditions or artificial obstructions should be borne by the employer, is sensible and logical. It is the employer's site; if he owns a difficult site, then the cost of dealing with the difficulties should fall upon him. Furthermore, the clause should ensure that the employer pays only for those difficulties that are encountered and not for the risk that they may be encountered where in the event they are not. However clause 12 alters the common law position – under the general law there is no warranty by the employer that the site is fit for the works: *Appleby* v. *Myers* (1867).

Nevertheless this clause, under earlier editions particularly, has been a prolific source of claims and disputes in which the contractor's claims were often, in the initial stages at least, met with rejection. The new wording of the Sixth Edition, and changes to clause 11 in particular, should promote the acceptance by employers of the need for a thorough subsoil investigation, and thereby achieve the objectives of reducing delay, cost and contention arising under clause 12.

The notices

Under the Sixth Edition the contractor is required to give notice to the engineer on encountering physical conditions or artificial obstructions which could not reasonably have been foreseen, whether or not he intends to claim. (Further, or simultaneous, notice is required of the intention to claim.) The notice must specify:

- The physical condition or artificial obstruction encountered.
- The expected effects thereof.
- The measures he is taking or proposing to take.
- The expected delay or interference with progress.

In many cases, the contractor is unable to give all of this information until some time after the condition or obstruction has been encountered. He should however provide the information as soon as it becomes available.

Although the notice requirements have been substantially relaxed from those imposed by earlier editions (in the Fourth Edition work carried out before giving notice could not be included in the claim), it should not be thought that they have become a mere formality. The notices, besides giving a warning to the engineer that a problem has arisen, provide him with an opportunity

- to require the contractor to investigate alternative measures for dealing with the conditions or obstruction;
- to vary the work so as to avoid the obstruction;
- to deal with the obstruction in some way different from that proposed by the contractor; or
- to stop progress on the work while he decides how to overcome the problem.

These options are available to the engineer under sub-clause (4).

Engineer's response

Where the engineer decides that the conditions or obstructions could not have been foreseen by an experienced contractor, he must determine the extension of time to which the contractor is entitled in respect of the delay caused, and must determine the payment to which the contractor is entitled, which in this case includes both costs and a reasonable percentage for profit.

Fair apportionment of risks

One of the main difficulties that arises in the administration of the clause is that the degree of probability of adverse conditions that the contractor is required to foresee is not defined. Clearly, an experienced contractor might foresee the *possibility* of many adverse physical conditions or obstructions to the work – is he required to allow for them all, however remote the possibility may be? If he does so and allows in his tender for the additional cost and delay, the employer will have to pay those additional costs whether or not the foreseen conditions or obstructions are actually encountered.

In *CJ Pearce & Co* v. *Hereford Corporation* (1986) the contractor had to excavate under a 100-year-old sewer known to exist at the time of making the contract, in order to lay a new sewer. The old sewer collapsed when the contractor disturbed the ground around it. The court held that the condition of the sewer could reasonably have been foreseen so that a claim under clause 12 would not have succeeded. A decision allowing a claim in such circumstances would have had the undesirable effect of implying a disincentive to take care when carrying out the work. A contractor who knows that the result of damaging an old sewer would be payment for the cost of repairing it plus a percentage of profit might be less than careful in his execution of the work around the sewer.

It is, however, reasonable to infer that the contractor is not expected to foresee and to allow for conditions or obstructions that are no more than mere possibilities. While his experience should tell him that remote possibilities exist, there can surely be no reason why he should provide for anything less than a substantial risk that the condition or obstruction may exist.

In many cases the engineer may himself foresee a substantial risk of an obstruction, and may safeguard against a clause 12 claim by instructing tenderers to allow in their rates for the risk or by providing provisional items in the bill of quantities to cover the foreseen risk. Conversely, a failure by the engineer to foresee such a risk may provide support for the contractor's contention he could not reasonably have been expected to foresee it.

The provision of sub-clause (5) for payment in respect of any variation ordered by the engineer before he reaches the decision that the conditions or obstructions were reasonably foreseeable, and hence that the clause 12 claim is invalid, may appear to be illogical. However, the decision of the engineer to order a variation would in many cases provide the contractor with a strong argument that the conditions could not reasonably have been foreseen. If an experienced contractor could have foreseen them, then so could the engineer; and had he done so, there should have been no need for a variation.

As a result of the decision of the Court of Appeal in *Holland Dredging (UK) Ltd* v. *The Dredging & Construction Co Ltd* (1987), it is quite clear that clause 12(1) is not limited to supervening events. Nor is clause 11(2) to be interpreted as meaning that the contractor bears the risk of unforeseen but existing adverse physical conditions and artificial obstructions, whether or not an experienced contractor would have reasonably foreseen them.

In *Humber Oil Terminals Trustee Ltd* v. *Harbour and General Works (Stevin) Ltd* (1991), the issue before the Court of Appeal was whether clause 12 of the Fifth Edition imposed liability on the employer for ground conditions that caused damage to the contractor's plant. The contractor was constructing mooring dolphins at Immingham Oil Terminal, using a jack-up barge equipped with a 300-tonne crane. While the crane was lifting a large precast soffit unit on to prepared piles, the barge listed and collapsed, causing extensive damage to both fixed and movable plant and equipment. The barge was a total loss and had to be replaced. There was much delay and extra cost. The contractor was held to be entitled to recover on the ground that the collapse of the barge and its consequences were due to encountering physical conditions that could not have been foreseen by an experienced contractor. Lord Justice Nourse said as follows:

> '[T]here is nothing to restrict the application of clause 12(1) to intransient, as distinct from transient, physical conditions. Indeed, the express reference to weather conditions, albeit by way of exclusion, suggests the contrary ... [while] I would agree that an applied stress is not of itself a physical condition, we are not concerned with such a thing in isolation, but with a combination of soil and an applied stress ... for the purposes of clause 12(1) you cannot dissociate the nature of the ground from an actual or notional application of some degree of stress. Without such an application you cannot predict how the ground will behave ... the condition encountered by the contractors was soil which behaved in an unforeseeable manner under the stress which was applied to it, and that was a physical condition within clause 12(1).'

The Court of Appeal also rejected the employer's argument that clause 8(2) – which imposes responsibility on the contractor for the adequacy, stability and safety for all site operations and methods of construction – imposed an unqualified full responsibility on the contractor. It did not apply to inadequacy or instability brought about by physical conditions within clause 12(1).

In general, the likelihood that clause 12 claims will arise is greatly reduced where there has been a comprehensive subsoil survey, the results of which have been included in the tender documents. The engineer can further minimise the risk of clause 12 claims arising by himself foreseeing likely obstructions and referring to them in the appropriate bill items. In law, the likelihood that a disclaimer as to the accuracy of subsoil information will be effective is debatable. Following the introduction of the words in parentheses in sub-clause (1) of clause 11 of the Fifth Edition and the new wording in the Sixth Edition, the likelihood of a successful defence to a clause 12 claim based upon clause 11 is further reduced.

These changes should not, however, be taken to be detrimental to the employer's interests. A contractor who knows that he will be recompensed for the cost of dealing with unforeseeable obstructions is able to price his work more accurately, and need not include for the cost of dealing with obstructions which may or may not exist. Hence the employer is likely to benefit from keener pricing by tenderers.

The likely consequences of an inadequate subsoil investigation, giving rise to a successful claim under clause 12, are:

- delayed completion of the works;
- additional costs resulting from the delay;
- additional costs of the unforeseen work resulting from lack of pre-planning and hence inefficient execution of the work; and
- additional costs of the unforeseen work arising from lack of competition in its evaluation.

Furthermore being unforeseen, neither the delay nor the additional expenditure would have been allowed for in the employer's budget.

3.5.3 Minor Works Conditions

Clause 3.8 provides:

(1) If during the carrying out of the Works the Contractor encounters any artificial obstruction or physical condition (other than a weather condition or a condition due to weather) which obstruction or condition

could not in his opinion reasonably have been foreseen by an experienced contractor the Contractor shall as early as practicable give written
notice thereof to the Engineer.

(2) If in the opinion of the Engineer such obstruction or condition could
not reasonably have been foreseen by an experienced contractor then the
Engineer shall certify a fair and reasonable sum and the Employer shall
pay such sum to cover the cost of performing any additional work or
using any additional plant or equipment together with a reasonable
percentage addition in respect of profit as a result of

(a) complying with any instructions which the Engineer may
issue and/or

(b) taking proper and reasonable measures to overcome or deal
with the obstruction or condition in the absence of instructions
from the Engineer

together with such sum as shall be agreed as the additional cost to the
Contractor of the delay or disruption arising therefrom. Failing agreement of such sums the Engineer shall determine the fair and reasonable
sum to be paid.

This is a simplified version of clause 12 of the ICE Conditions, and it
is similar in its approach. Here, however, the contractor's duty is
merely to give written notice of the artificial obstruction or physical
condition to the engineer 'as early as practicable'. If the engineer
forms the opinion that the obstruction or condition could not have
been reasonably foreseen by an experienced contractor, he is to
determine and certify a fair and reasonable sum (including a
reasonable percentage for profit) to cover the cost of carrying out
any necessary additional work, etc. required as a result of the
contractor's compliance with any instructions given by the engineer
to overcome the problem *and/or* the contractor having taken proper
and reasonable measures to deal with the problem in the absence of
the engineer's instruction.

In addition, the contractor is entitled to 'such sum as shall be
agreed as the additional cost to [him] of' any resultant delay or
disruption. Failing agreement, it is for the engineer to determine a
fair and reasonable amount. If, as a result, there is delay to progress,
there will also be entitlement to extension of time under clause
4.4(c).

3.6 *Satisfaction of the engineer*

Clause 13 of the ICE Conditions states:

(1)   Save insofar as it is legally or physically impossible the Contractor shall construct and complete the Works in strict accordance with the Contract to the satisfaction of the Engineer and shall comply with and adhere strictly to the Engineer's instructions on any matter connected therewith (whether mentioned in the Contract or not). The Contractor shall take instructions only from the Engineer or (subject to the limitations referred to in Clause 2) from the Engineer's Representative.

(2)   The whole of the materials Contractor's Equipment and labour to be provided by the Contractor under Clause 8 and the mode manner and speed of construction of the Works are to be of a kind and conducted in a manner acceptable to the Engineer.

(3)   If in pursuance of Clause 5 or sub-clause (1) of this Clause the Engineer shall issue instructions which involve the Contractor in delay or disrupt his arrangements or methods of construction so as to cause him to incur cost beyond that reasonably to have been foreseen by an experienced contractor at the time of tender then the Engineer shall take such delay into account in determining any extension of time to which the Contractor is entitled under Clause 44 and the Contractor shall subject to Clause 52(4) be paid in accordance with Clause 60 the amount of such cost as may be reasonable except to the extent that such delay and extra cost result from the Contractor's default. Profit shall be added thereto in respect of any additional permanent or temporary work. If such instructions require any variation to any part of the Works the same shall be deemed to have been given pursuant to Clause 51.

Legal impossibility covers cases where, for example, the work cannot be performed without infringement of planning law or building regulations, or where a process or material to be used is protected by patent and no licence can be obtained. Physical impossibility could cover a situation where the contractor is required to drive piles to a specified penetration but is prevented from doing so by the presence of rock. In both cases, the remedy is for the engineer to order a variation. It is, however, implicit in the clause that the contractor was not and could not have been aware of the impossibility at the time of making the contract. Accordingly, a physical impossibility may form the subject of a claim under clause 12.

Where the impossibility arises from some event that occurred after the contract was made – for example, the destruction by fire of a building that was to have been extended – then the position is dealt with under clause 64 (frustration).

The consequences of instructions issued by the engineer in order to avoid the impossibility may be delay and/or disruption, in respect of which the contractor is entitled to an appropriate extension of time and to reimbursement of cost. Where a variation order is issued, the value of that variation is determined in accordance with clause 52. Clause 52(4) is of course the general requirement of notice in respect of claims.

Sub-clause (1) confirms that, subject only to relief in a case of legal or physical impossibility, the contractor has an absolute obligation to construct and complete the works in strict accordance with the contract. The obligation should be considered carefully by a contractor who is contemplating stopping work because of some alleged breach of the contract by the employer. Unless he is able to prove that breach, and that it is a fundamental breach, going to the root of the contract, then it is the contractor who will be in breach and who will have to bear the losses, both his own and those of the employer, that flow from that breach.

Sub-clauses (2) and (3) give the engineer power to approve or to disapprove the materials, methods and workmanship of the contractor – impliedly whether or not they are in accordance with the contract. Sub-clause (3) does, however, provide for reimbursement to the contractor of any cost beyond that reasonably to have been foreseen by an experienced contractor in complying with the engineer's instructions or directions under the clause, together with an addition thereto for profit. Clearly, the contractor should have foreseen any instruction or direction that merely requires compliance with the specification, drawings, etc. Hence a claim will arise only where the contractor can show that the engineer's requirements exceed those of the contract.

There is no corresponding clause under the Minor Works Form.

3.7 Construction programme

3.7.1 The ICE Conditions

Clause 14 of the Sixth Edition states:

(1) (a) Within 21 days after the award of the Contract the Contractor shall submit to the Engineer for his acceptance a programme showing the order in which he proposes to carry out the Works having regard to the provisions of Clause 42(1).

(b) At the same time the Contractor shall also provide in writing for the information of the Engineer a general description of the arrangements and methods of construction which the Contractor proposes to adopt for the carrying out of the Works.

61

(c)   Should the Engineer reject any programme under sub-clause (2)(b) of this Clause the Contractor shall within 21 days of such rejection submit a revised programme.

(2)   The Engineer shall within 21 days after receipt of the Contractor's programme

(a)   accept the programme in writing or

(b)   reject the programme in writing with reasons or

(c)   request the contractor to supply further information to clarify or substantiate the programme or to satisfy the Engineer as to its reasonableness having regard to the Contractor's obligations under the Contract.

Provided that if none of the above actions is taken within the said period of 21 days the Engineer shall be deemed to have accepted the programme as submitted.

(3)   The Contractor shall within 21 days after receiving from the Engineer any request under sub-clause (2)(c) of this Clause or within such further period as the Engineer may allow provide the further information requested failing which the relevant programme shall be deemed to be rejected. Upon receipt of such further information the Engineer shall within a further 21 days accept or reject the programme in accordance with sub-clauses (2)(a) or (2)(b) of this Clause.

(4)   Should it appear to the Engineer at any time that the actual progress of the work does not conform with the accepted programme referred to in sub-clause (1) of this Clause the Engineer shall be entitled to require the Contractor to produce a revised programme showing such modifications to the original programme as may be necessary to ensure completion of the Works or any Section within the time for completion as defined in Clause 43 or extended time granted pursuant to Clause 44. In such event the Contractor shall submit his revised programme within 21 days or within such further period as the Engineer may allow. Thereafter the provisions of sub-clauses (2) and (3) of this Clause shall apply.

(5)   The Engineer shall provide to the Contractor such design criteria relevant to the Permanent Works or any Temporary Works design supplied by the Engineer as may be necessary to enable the Contractor to comply with sub-clauses (6) and (7) of this Clause.

(6)   If requested by the Engineer the Contractor shall submit at such times and in such further detail as the Engineer may reasonably require information pertaining to the methods of construction (including Temporary Works and the use of Contractor's Equipment) which the Contractor proposes to adopt or use and calculations of stresses strains and deflections that will arise in the Permanent Works or any parts thereof during construction so as to enable the Engineer to decide whether if these methods are adhered to the Works can be constructed and completed in accordance with the Contract and without detriment to the Permanent Works when completed.

(7) The Engineer shall inform the Contractor in writing within 21 days after receipt of the information submitted in accordance with sub-clauses (1)(b) and (6) of this Clause either

(a) that the Contractor's proposed methods have the consent of the Engineer or

(b) in what respects in the opinion of the Engineer they fail to meet the requirements of the Contract or will be detrimental to the Permanent Works.

In the latter event the Contractor shall take such steps or make such changes in the said methods as may be necessary to meet the Engineer's requirements and to obtain his consent. The Contractor shall not change the methods which have received the Engineer's consent without the further consent in writing of the Engineer which shall not be unreasonably withheld.

(8) If the Contractor unavoidably incurs delay or cost because

(a) the Engineer's consent to the proposed methods of construction is unreasonably delayed or

(b) the Engineer's requirements pursuant to sub-clause (7) of this Clause or any limitations imposed by any of the design criteria supplied by the Engineer pursuant to sub-clause (5) of this Clause could not reasonably have been foreseen by an experienced contractor at the time of tender

the Engineer shall take such delay into account in determining any extension of time to which the Contractor is entitled under Clause 44 and the Contractor shall subject to Clause 52(4) be paid in accordance with Clause 60 such sum in respect of the cost incurred as the Engineer considers fair in all the circumstances. Profit shall be added thereto in respect of any additional permanent or temporary work.

(9) Acceptance by the Engineer of the Contractor's programme in accordance with sub-clauses (2) (3) or (4) of this Clause and the consent of the Engineer to the Contractor's proposed methods of construction in accordance with sub-clause (7) of this Clause shall not relieve the Contractor of any of his duties or responsibilities under the Contract.

The contractor's programme, although required under sub-clause (1) to show only the order of procedure, impliedly must also show the time required to perform the operations indicated therein. Without a timescale, it would not be possible for the engineer to determine under sub-clause (4) of this clause or under clause 46 whether or not the actual progress of the works conforms with the approved programme.

3.7.2 Programmed early completion

A question sometimes arises as to the validity of a contractor's clause 14 programme which shows completion in substantially less time than he is allowed under the contract. Such a programme may well be proposed by a contractor who finds the available time to be over-generous: for by completing early he will save both site and head-office overheads. Although one might expect the employer to favour such a programme because the works will be available to him earlier than he had expected, that is not always the case.

There is a popular belief held by some employers or by their engineers that the purpose of such a programme is to pave the way for claims that might not otherwise exist. This belief is presumably founded on the erroneous theory that the very fact of completing late entitles the contractor to claim for delay: hence that if he takes the full time available to him under the contract, having planned to take, for example, three months less than that time, then he will automatically be able to claim that he has been delayed by three months, for which delay the employer is responsible.

The true position is, of course, that delay claims arise only where the contractor can show that he has been delayed by some cause for which the employer is responsible. Moreover, where the contractor's clause 14 programme shows completion before the time specified in the contract it is to be inferred from *Glenlion Construction* v. *The Guinness Trust* (1987) (which arose from a contract under the JCT Standard Form of Building Contract) that the contractor will not have a claim for delay unless his completion is delayed until after the *contractual* time for completion has expired.

There are other possible reasons why the engineer may wish the contractor to proceed at a more leisurely pace than is proposed in a programme providing for early completion. If the engineer foresees difficulty in completing the working drawings within the time available or if his employer has budgeted for a slower rate of expenditure than may be required under the contractor's proposals, then either or both may be in difficulty. However in neither case can the engineer disapprove a programme on such grounds, unless of course limits to rates of expenditure or to rates of progress of the works have been set down in the contract documents. The requirement of the contract is that the contractor shall complete the works *within* the time prescribed by that contract.

Should it appear to the engineer that the contractor's programme of works (in relation to his methods of construction and his intentions as to plant and labour to be employed) is unrealistic, then he can and should require the further details and information that may be necessary to satisfy himself as to the adequacy of the contractor's proposals. He is entitled to this under sub-clauses (1) and (2).

3.7.3 Revised programmes

Where revised programmes become necessary under sub-clause (4) it is important that a dated copy of every such programme is retained by the engineer and by the contractor as evidence of the contractor's intentions as at that date.

In sub-clause (7) the word 'consent' in relation to the contractor's proposed methods was chosen to distinguish his action from the 'approval' given to some other matters in the contract. The purpose is to indicate that such consent does not make the engineer liable for, for example, the contractor's design of temporary works. This is, however, made clear in sub-clause (7)(a).

Claims in respect of delay and for additional costs incurred plus profit thereon may arise where the engineer's consent is unreasonably delayed or where his requirements could not reasonably have been foreseen by an experienced contractor. Here again, the experienced contractor should have foreseen that his methods must comply with any specified requirement, so that in many cases claims under the clause may be avoided where the engineer provides, as part of the tender documents, full information of any design criteria that may affect construction methods.

3.7.4 Incorporation of programme into the contract

The programming requirement has clearly been drafted to facilitate the administration of the contract and the execution of the works, but it is not surprising that difficulties can and do arise. A case in point is *Yorkshire Water Authority* v. *Sir Alfred McAlpine & Son (Northern) Ltd* (1985), which arose out of a contract which incorporated the Fifth Edition of the ICE Conditions. The specification required the contractor to submit with his tender a programme showing that he had taken note of certain specified phasing requirements, this being 'in addition to the requirements of clause

14'. His method statement as submitted was approved and was incorporated in the contract. It provided for the construction of certain tunnelling works upstream. This in fact proved impossible and the contractor proceeded with the work downstream and claimed a variation under clause 51(1) with its reference to 'changes in the specified sequence method or timing of construction'.

The High Court ruled in the contractor's favour. The method statement was not the clause 14 programme. Its incorporation into the contract imposed an obligation on the contractor to follow it, save in so far as it was legally or physically impossible to do so. The method statement had become a specified method of construction and the contractor was entitled to a variation order and payment accordingly. As Mr Justice Skinner aptly put it:

'The standard conditions recognise a clear distinction between obligations specified in the contract in detail, which both parties can take into account when agreeing a price, and those which are general and which do not have to be specified precontractually. In this case [the employers] could have left the programme and methods as the sole responsibility of the [contractors] under clause 14(1) and clause 14(3). The risks inherent in such a programme or method would then have been the [contractors'] throughout. Instead they decided they wanted more control over the methods and programme than clause 14 provided. Hence ... the method statement, [and] the incorporation of the method statement into the contract imposing the obligation on the [contractors] to follow it save in so far as it was legally or physically impossible to do so. It therefore became a specified method of construction by agreement between the parties.'

3.7.5 Minor Works Conditions

The programme requirement in the Minor Works Conditions is much simpler. It is provided for in clause 4.3:

The Contractor shall within 14 days after the starting date if so required provide a programme of his intended activities. The Contractor shall at all times proceed with the Works with due expedition and reasonably in accordance with his programme or any modification thereof which he may provide or which the Engineer may request.

The contractual requirement here is to provide a programme of the contractor's intended activities if so required. The contractor must then proceed with the works *inter alia* 'reasonably in accordance with his programme or any modification thereof which he may provide or which the Engineer may request'. If the engineer does request a modification of the programme, there is no corresponding contractual right to time or money. This is something which is contemplated by the contracting parties, and the potential risk to the contractor is small.

3.8 Engineer's setting-out errors

Clause 17 provides:

(1)   The Contractor shall be responsible for the true and proper setting-out of the Works and for the correctness of the position levels dimensions and alignment of all parts of the Works and for the provision of all necessary instruments appliances and labour in connection therewith.

(2)   If at any time during the progress of the Works any error shall appear or arise in the position levels dimensions or alignment of any part of the Works the Contractor on being required so to do by the Engineer shall at his own cost rectify such error to the satisfaction of the Engineer unless such error is based on incorrect data supplied in writing by the Engineer or the Engineer's Representative in which case the cost of rectifying the same shall be borne by the Employer.

(3)   The checking of any setting-out or of any line or level by the Engineer or the Engineer's Representative shall not in any way relieve the Contractor of his responsibility for the correctness thereof and the Contractor shall carefully protect and preserve all bench-marks sight rails pegs and other things used in setting out the Works.

Responsibility for the cost of dealing with setting-out errors is clearly defined in clause 17. Where the errors originate from incorrect data supplied in writing by the engineer or by the resident engineer, the employer must bear the cost. In other cases, the contractor must carry out whatever remedial work may be needed at his own expense.

The wording of the clause appears to indicate that the supply of setting out data is within the authority of the engineer's representative, whether or not it is expressly delegated to him under clause 2(4).

There are occasions upon which the engineer's representative may set out some part of the works himself. For example, he may drive pegs into the ground to indicate centre lines of roads or of manholes. Setting-out data supplied in this way is clearly not 'in writing' unless the contractor is also provided with a written instruction that he should take the pegs to indicate the location of the relevant centre lines. In such cases, it is suggested that the note should include dimensions from the pegs to fixed objects that will not be disturbed by the works, sufficient to determine the positions of the pegs during and after construction of the works. Setting-out pegs or other markers are often damaged or moved or destroyed during the course of the works, as everyone is well aware.

Where the contractor asserts that a setting-out error was caused by incorrect data supplied by the engineer or his representative, the burden of proof lies on the contractor. Hence it is in the contractor's interests to ensure that any physical setting-out by the engineer or his representative is accompanied by a written instruction enabling the location of the markers used to be determined at a later date. Where necessary, the contractor should adopt the clause 51(2) procedure of seeking written confirmation of an oral instruction so as to provide himself with evidence should a dispute arise.

Under clause 3.5 of the Minor Works Conditions, setting out is the contractor's responsibility.

3.9 Boreholes and exploratory excavation

Clause 18 of the ICE Conditions states:

> If at any time during the construction of the Works the Engineer shall require the Contractor to make boreholes or to carry out exploratory excavation such requirement shall be ordered in writing and shall be deemed to be a variation ordered under Clause 51 unless a Provisional Sum or Prime Cost Item in respect of such anticipated work shall have been included in the Bill of Quantities.

The need for unforeseen site investigation sometimes arises under contracts carried out under the ICE Conditions. Subsoil conditions at the intended founding level may, for example, be found to be substantially different from those upon which the design is based. In such cases, the engineer may decide that boreholes are needed to check the strata below that level, both to find the depth to a suitable stratum and to ensure that ground below that stratum is suitable. An instruction to carry out such investigation is, logically, deemed to be a variation under clause 51. In turn, this clause provides for the value of the work, determined impliedly in accordance with clause 52, to be taken into account in the contract price.

There is no corresponding provision under the Minor Works Form.

3.10 *Damage occasioned by excepted risks*

3.10.1 ICE Conditions

Clause 20 of the ICE Conditions states:

(1) (a) The Contractor shall save as in paragraph (b) hereof and subject to sub-clause (2) of this Clause take full responsibility for the care of the Works and materials plant and equipment for incorporation therein from the Works Commencement Date until the date of issue of a Certificate of Substantial Completion for the whole of the Works when the responsibility for the said care shall pass to the Employer.

(b) If the Engineer issues a Certificate of Substantial Completion for any Section or part of the Permanent Works the Contractor shall cease to be responsible for the care of that Section or part from the date of issue of such Certificate of Substantial Completion when the responsibility for the care of that Section or part shall pass to the Employer.

(c) The Contractor shall take full responsibility for the care of any outstanding work and materials plant and equipment for incorporation therein which he undertakes to finish during the Defects Correction Period until such outstanding work has been completed.

(2) The Excepted Risks for which the Contractor is not liable are loss or damage to the extent that it is due to

(a) the use or occupation by the Employer his agents servants or other contractors (not being employed by the Contractor) of any part of the Permanent Works

(b) any fault defect error or omission in the design of the Works (other than a design provided by the Contractor pursuant to his obligations under the Contract)

(c) riot war invasion act of foreign enemies or hostilities (whether war be declared or not)

(d) civil war rebellion revolution insurrection or military or usurped power

(e)   ionizing radiations or contamination by radioactivity from any nuclear fuel or from any nuclear waste from the combustion of nuclear fuel radioactive toxic explosive or other hazardous properties of any explosive nuclear assembly or nuclear component thereof and

(f)   pressure waves caused by aircraft or other aerial devices travelling at sonic or supersonic speeds.

(3)   (a)   In the event of any loss or damage to

(i)   the Works or any Section or part thereof or

(ii)   materials plant or equipment for incorporation therein

while the Contractor is responsible for the care thereof (except as provided in sub-clause (2) of this Clause) the Contractor shall at his own cost rectify such loss or damage so that the Permanent Works conform in every respect with the provisions of the Contract and the Engineer's instructions. The Contractor shall also be liable for any loss or damage to the Works occasioned by him in the course of any operations carried out by him for the purpose of complying with his obligations under Clauses 49 and 50.

(b)   Should any such loss or damage arise from any of the Excepted Risks defined in sub-clause (2) of this Clause the con-tractor shall if and to the extent required by the Engineer rectify the loss or damage at the expense of the Employer.

(c)   In the event of loss or damage arising from an Excepted Risk and a risk for which the Contractor is responsible under sub-clause (1)(a) of this Clause then the Engineer shall when deter-mining the expense to be borne by the Employer under the Con-tract apportion the cost of rectification into that part caused by the Excepted Risk and that part which is the responsibility of the Contractor.

Whether or not the contractor insures against the risk of damage to the works, he remains liable to the employer in respect of any such damage. His liability continues until the issue of the certificate of completion. The contractor is not however liable for loss or damage caused by one of the 'excepted risks' defined in sub-clause (2).

Of the excepted risks, 'fault defect error or omission in the design of the works' is probably the most prolific source of claims. 'Design' in this context impliedly includes the specification and any other document issued by the engineer, either as part of the contract documents or during the course of the works since clearly addi-tional or varied drawings and variation orders must constitute part of the design. The burden of proving that the damage was caused by an 'excepted risk' lies on the contractor.

Unlike some other contractual provisions, clause 20 does not make specific provision for delay and disruption costs occasioned by design faults. It simply provides that remedial works necessary to deal with such a fault (and damage caused by the other excepted risks) shall be at the 'expense' of the employer.

The use of the word 'expense' instead of the term 'cost' is significant. The contract contemplates that under its express terms the contractor's entitlement is to the rectification costs. (See section 3.1 above.)

In many cases of course reimbursement of 'cost' as defined in clause 1(5) will provide adequate compensation for the losses resulting from the need to correct a design fault, since additional site and head-office overheads resulting from the remedial works will be recovered within the definition of 'cost'. Where the design fault has to be investigated and a remedy has to be found before remedial work may proceed, it is to be expected that the engineer will or should issue a suspension order under clause 40. This will itself entitle the contractor to reimbursement of his additional costs consequent upon the delay.

3.10.2 Minor Works Conditions

Under the Minor Works Conditions, the position is similar. Clause 3.3 provides:

> (1)   In case any damage loss or injury from any cause whatsoever (save and except the Excepted Risks) shall happen to the Works or any part thereof while the Contractor is responsible for their care the Contractor shall at his own cost repair and make good the same so that at completion the Works shall be in good order and condition and conform in every respect with the requirements of the Contract and the Engineer's instructions.
> (2)   To the extent that any damage loss or injury arises from any of the Excepted Risks the Contractor shall if required by the Engineer repair and make good the same at the expense of the Employer.
> (3)   The Contractor shall also be liable for any damage to the Works occasioned by him in the course of any operations carried out by him for the purpose of completing outstanding work or complying with his obligations under Clauses 4.7 and 5.2.

There is provision (clause 10.1), at the option of the employer, for the contractor to maintain insurance in the joint names of employer and contractor to cover both the permanent and temporary works to their full value, and the constructional plant to its full value, against all loss or damage. The only exception is loss or damage caused by 'excepted risks'.

The use of the words in clause 3.3(1) 'from any cause whatsoever' makes it clear that there is no implied limitation to the contractor's responsibility, even if it is outside the contemplation of the parties, and (save for damage caused by the 'excepted risks') even if it arises from the employer's own negligence. This follows from the principle established in *AE Farr Ltd* v. *The Admiralty* (1953) as a matter of interpretation of the contract. Even though the contract was not in ICE form, the term at issue in that case was identically phrased and Mr Justice Parker, discussing the meaning of the term 'any cause whatsoever', said:

> 'It seems to me that the words ... are about as wide as they can be and I must read 'any cause whatsoever' as if it included, and expressly said, 'including damage caused by the negligent navigation by an Admiralty servant of a ship'.

Both ICE conditions clause 20(3) and the Minor Works Conditions clause 3.3 make the contractor responsible for damage caused and he must make it good. The problem is theoretical rather than real because in practice most contractors carry all-risks insurance. Difficulties will only arise if contractors carry large excesses.

3.11 New Roads and Street Works Act 1991

3.11.1 ICE Conditions

Clause 27 of the ICE Conditions states:

New Roads and Street Works Act 1991 – Definitions
(1) (a) In this Clause 'the Act' shall mean the New Roads and Street Works Act 1991 and any statutory modification or re-enactment thereof for the time being in force.
(b) For the purpose of obtaining any licence under the Act required for the permanent works the undertaker shall be the Employer who for the purposes of the Act will be the licensee.
(c) For all other purposes the undertaker under the licence shall be the Contractor.
(d) All other expressions common to the Act and to this Clause shall have the same meaning as those assigned to them by the Act.

Licences

(2) (a) The Employer shall obtain any street works licence and any other consent licence or permission that may be required for the carrying out of the Permanent Works and shall supply the Contractor with copies thereof including details of any conditions or limitations imposed.

(b) Any condition or limitation in any licence obtained after the award of the Contract shall be deemed to be an instruction under Clause 13.

Notices

(3) The Contractor shall be responsible for giving to any relevant authority any required notice (or advance notice where prescribed) of his proposal to commence any work. A copy of each such notice shall be given to the Employer.

The New Roads and Street Works Act 1991 ('the 1991 Act') repealed the Public Utilities Street Works Act 1950. It was implemented in stages and became fully operational from 1 January 1993. The 1991 Act contains complex and detailed provisions that are intended to provide street authorities with control over the installation of underground apparatus (by statutory water, sewerage, electricity, gas, and telephone authorities) beneath roads and beneath ground upon which roads are intended to be constructed. The legislation has resulted in similarly complex and detailed provisions in the ICE Conditions, of which the main requirement from the contractor's viewpoint is that notice is required of intention to commence work in land referred to in the 1991 Act. The notice period is related to the type of works and the type of 'street', which term includes 'land laid out as a way'.

Clause 27 imposes an obligation on the employer to be 'the undertaker' for the purpose of obtaining any licences or other consents that may be required in connection with the permanent works. Upon receipt of an application for a licence it is a responsibility of the street authority to give notice of the proposed works to other authorities or persons having apparatus in the street likely to be affected by the works. Where the street authority so directs 'advance notice' of certain works may be required, and where such notice is prescribed the undertaker may be required to provide additional information needed in order to co-ordinate the proposed works with 'other works of any description proposed to be executed in the street'.

Although it is clear that compliance with the 1991 Act in respect of obtaining licences may be time-consuming, it is to be expected that the employer will, through his engineer, ensure that all necessary licences are obtained well in advance of making the contract; thereby avoiding delay to the construction of the works. Where this aim is not achieved however – for example where variations requiring a varied licence become necessary – additional delays are likely to result. Such delays, and the costs associated with them, entitle the contractor to claim under sub-clause (2)(b) under which any condition or limitation in a licence obtained after the award of the contract is deemed to be an instruction given under clause 13.

For the purpose of notices of the starting date of the works the contractor becomes 'the undertaker' and as such is responsible for giving notice both to the street authority and to any other authority or person having apparatus in the street likely to be affected by the works. Although such notice may be as little as seven working days, the street authority may prescribe a longer period. The notice becomes invalid if work is not commenced within seven working days of the specified commencement date.

3.11.2 Minor Works Conditions

No express provision appears in the Minor Works form in respect of the 1991 Act. Statutory obligations are dealt with generally under clause 9, which obliges the contractor to comply with all statutory requirements, including giving notices and paying fees. Sub-clause 9.2 makes the employer responsible for obtaining consents. Under sub-clause 9.3 the contractor is not liable to the employer for any failure to comply with statutory requirements resulting from his carrying out the works in accordance with the contract or any engineer's instruction.

3.12 *Provision of facilities for other contractors*

3.12.1 ICE Conditions

Clause 31 of the ICE Conditions states:

(1) The Contractor shall in accordance with the requirements of the Engineer or Engineer's Representative afford all reasonable facilities for

any other contractors employed by the Employer and their workmen and for the workmen of the Employer and of any other properly authorised authorities or statutory bodies who may be employed in the execution on or near the Site of any work not in the Contract or of any contract which the Employer may enter into in connection with or ancillary to the Works.

(2)   If compliance with sub-clause (1) of this Clause shall involve the Contractor in delay or cost beyond that reasonably to be foreseen by an experienced contractor at the time of tender then the Engineer shall take such delay into account in determining any extension of time to which the Contractor is entitled under Clause 44 and the Contractor shall subject to Clause 52(4) be paid in accordance with Clause 60 the amount of such cost as may be reasonable. Profit shall be added thereto in respect of any additional permanent or temporary work.

In the Sixth Edition, the RE is empowered to require the contractor to provide facilities for other contractors and impliedly the exercise of this power does not depend on formal delegation under clause 2(4). An innovation in the Sixth Edition is the inclusion of profit in the contractor's entitlement in respect of costs incurred for additional permanent or temporary work.

The requirement to provide 'all reasonable facilities' for any other contractors employed by the employer may appear to be onerous, especially where such facilities include access roads to the site, scaffolding, and the site itself, which may already be congested. But the contractor is not entitled to exclusive possession of the site and the inclusion of the word 'reasonable', and provisions for extension of time and reimbursement of cost beyond that reasonably to be foreseen, will in most cases provide the contractor with adequate recompense for delay or additional costs incurred.

For example, if other contractors' proposed use of access roads causes delay to the contractor's own vehicles, or damage to the road, it would appear that the contractor could assert that the requirement is not reasonable and refuse the engineer's instruction. Alternatively, of course, he could agree an appropriate price in advance. Where the contractor is required to leave scaffolding in position for a longer period than was provided for in his tender, he should be able to claim reimbursement of the additional costs incurred.

'Reasonable' is an important qualification and it is suggested that if the facilities required would seriously disrupt the contractor's programme they are not within the clause.

If the contractor has to provide, in response to the engineer's requirements, facilities that are not 'reasonable' he would then have a claim under clause 31(2) or clause 51. Good practice dictates that wherever possible the engineer should foresee any likely requirements by other contractors and provide details with the tender documents, thereby obviating a claim.

3.12.2 Minor Works Conditions

Clause 3.9 is the corresponding provision under the Minor Works contract. It provides:

> 3.9 The Contractor shall in accordance with the requirements of the Engineer afford reasonable facilities for any other contractor employed by the Employer and for any other properly authorized authority employed on the Site.

The effect of this provision is the same as that of clause 31, except that there is no reference to delay and extra cost. Delay would be dealt with by an extension of time under clause 4.4(h), but it is thought that the contractor would have no financial claim except to the extent that the engineer's requirements can be shown to be unreasonable.

3.13 *Disposal of fossils etc.*

Clause 32 of the ICE Conditions states:

> All fossils coins articles of value or antiquity and structures or other remains or things of geological or archaeological interest discovered on the Site shall as between the Employer and the Contractor be deemed to be the absolute property of the Employer and the Contractor shall take reasonable precautions to prevent his workmen or any other persons from removing or damaging any such article or thing and shall immediately upon discovery thereof and before removal acquaint the Engineer of such discovery and carry out at the expense of the Employer the Engineer's orders as to the disposal of the same.

It is clear from the definition of 'site' in clause 1 that in this context it must include borrow pits, etc. that are away from the construction site where such pits are provided for under the contract.

Clause 32 clarifies the right of the employer to take possession of any articles of value that may be found on the site, although he may himself have to hand them over to the Crown where they are deemed to be 'treasure trove'. Treasure trove is gold or silver coin, plate, bullion or other valuable items *hidden* in the earth or other secret place, the true owner being unknown. A coroner's inquest is held to determine whether or not the objects are treasure trove. From the contractor's viewpoint the problem that often arises from the clause is that of delay; where old foundations or objects of archaeological interest are discovered and careful uncovering becomes necessary in order to preserve the relics.

Here, as in clause 20, the word 'expense' seems to have been used instead of 'cost' to mean something more than that term as defined. In other words, the contractor is entitled to his costs and a reasonable profit. Although not explicit in the clause, it is submitted that additional costs resulting from delay to the works occasioned by compliance with the clause must be included in the reference to 'at the expense of the employer'. Any delay would be dealt with by an extension of time under clause 44.

There is no reference to fossils and articles of value in the Minor Works form, but there is a right to extension of time under clause 4.4(h). Clearly if such objects are found the contractor should notify the engineer and seek an instruction, which might rank for payment. The contractor may also have a remedy under clause 3.8.

3.14 *Cost of samples and of tests*

Clause 36 of the ICE Conditions states:

(1)   All materials and workmanship shall be of the respective kinds described in the Contract and in accordance with the Engineer's instructions and shall be subjected from time to time to such tests as the Engineer may direct at the place of manufacture or fabrication or on the Site or such other place or places as may be specified in the Contract. The Contractor shall provide such assistance instruments machines labour and materials as are normally required for examining measuring and testing any work and the quality weight or quantity of any materials used and shall supply samples of materials before incorporation in the Works for testing as may be selected and required by the Engineer.
(2)   All samples shall be supplied by the Contractor at his own cost if the supply thereof is clearly intended by or provided for in the Contract but if not then at the cost of the Employer.
(3)   The cost of making any test shall be borne by the Contractor if such

test is clearly intended by or provided for in the Contract and (in the cases only of a test under load or of a test to ascertain whether the design of any finished or partially finished work is appropriate for the purposes which it was intended to fulfil) is particularized in the Specification or Bill of Quantities in sufficient detail to enable the Contractor to have priced or allowed for the same in his Tender. If any test is ordered by the Engineer which is either

(a)   not so intended by or provided for or

(b)   (in the cases above mentioned) is not so particularized

then the cost of such test shall be borne by the Contractor if the test shows the workmanship or materials not to be in accordance with the provisions of the Contract or the Engineer's instructions but otherwise by the Employer.

Where, as for example in the case of a sewer, routine tests on each section laid are specified in the contract, the engineer is entitled to re-tests at no additional cost to the employer where any test indicates failure. Similarly, additional tests at no extra cost to the employer can impliedly be ordered where tests on cubes of concrete or samples of any other material fail to comply with the specification.

The need to order tests in addition to those provided for in the contract usually arises because testing has been omitted from the contract in error, or because some event during construction throws doubt on the suitability of the material in question. Additional tests may be ordered only at the place of manufacture or fabrication or on the site since any other location must be specified in the contract. In such cases, the contractor's right to payment depends upon the test showing the sample to be in accordance with the specification. The cost is to be borne by the contractor only 'if the test shows the workmanship or materials not to be in accordance with the con-tract'.

Clause 2.3(b) of the Minor Works contract empowers the engineer to order the 'carrying out of any test or investigation'. If the testing shows compliance with the contract, the contractor has a right to extension of time (clause 4.4(b)) and to additional cost (clause 6.1).

3.15   Cost of uncovering work

Clause 38 of the ICE Conditions states:

(1)   No work shall be covered up or put out of view without the consent of the Engineer and the Contractor shall afford full opportunity for the

Engineer to examine and measure any work which is about to be covered up or put out of view and to examine foundations before permanent work is placed thereon. The Contractor shall give due notice to the Engineer whenever any such work or foundations is or are ready or about to be ready for examination and the Engineer shall without unreasonable delay unless he considers it unnecessary and advises the Contractor accordingly attend for the purpose of examining and measuring such work or of examining such foundations.
(2)   The Contractor shall uncover any part or parts of the Works or make openings in or through the same as the Engineer may from time to time direct and shall reinstate and make good such part or parts to the satisfaction of the Engineer. If any such part or parts have been covered up or put out of view after compliance with the requirements of sub-clause (1) of this Clause and are found to be executed in accordance with the Contract the cost of uncovering making openings in or through reinstating and making good the same shall be borne by the Employer but in any other case all such cost shall be borne by the Contractor.

It is implied that the engineer will not cause unreasonable delay in approving work after the contractor has notified him that it is ready for inspection before being covered up. If he does so, then the contractor is entitled to an extension of time in respect of the delay and to the costs resulting from that delay.

The right of the contractor to claim the cost of uncovering work depends upon two things.

- Firstly, he must have given notice to the engineer and obtained his approval before covering up.
- Secondly, the work, when uncovered, must prove to be without defect.

In such cases, it is submitted, the contractor is also able to claim an extension of time and consequential costs in respect of the delay caused by the instruction. The reference in clause 38(2) to 'cost . . . borne by the Employer' means, it is thought, that the contractor recovers the ordinary cost together with a reasonable percentage addition for profit.

In the Minor Works contract there is no corresponding provision, but there is no reason why a similar requirement should not be written into the contract documents.

3.16   Suspension of work

3.16.1   ICE Conditions

Clause 40 states:

(1)   The Contractor shall on the written order of the Engineer suspend the progress of the Works or any part thereof for such time or times and in such manner as the Engineer may consider necessary and shall during such suspension properly protect and secure the work so far as is necessary in the opinion of the Engineer. Subject to Clause 52(4) the Contractor shall be paid in accordance with Clause 60 the extra cost (if any) incurred in giving effect to the Engineer's instructions under this Clause except to the extent that such suspension is
(a)   otherwise provided for in the Contract or
(b)   necessary by reason of weather conditions or by some default on the part of the Contractor or
(c)   necessary for the proper execution or for the safety of the Works or any part thereof in as much as such necessity does not arise from any act or default of the Engineer or the Employer or from any of the Excepted Risks defined in Clause 20(2).
Profit shall be added thereto in respect of any additional permanent or temporary work. The Engineer shall take any delay occasioned by a suspension ordered under this Clause (including that arising from any act or default of the Engineer or the Employer) into account in determining any extension of time to which the Contractor is entitled under Clause 44 except when such suspension is otherwise provided for in the Contract or is necessary by reason of some default on the part of the Contractor.
(2)   If the progress of the Works or any part thereof is suspended on the written order of the Engineer and if permission to resume work is not given by the Engineer within a period of 3 months from the date of suspension then the Contractor may unless such suspension is otherwise provided for in the Contract or continues to be necessary by reason of some default on the part of the Contractor serve a written notice on the Engineer requiring permission within 28 days from the receipt of such notice to proceed with the Works or that part thereof in regard to which progress is suspended. If within the said 28 days the Engineer does not grant such permission the Contractor by a further written notice so served may (but is not bound to) elect to treat the suspension where it affects part only of the Works as an omission of such part under Clause 51 or where it affects the whole Works as an abandonment of the Contract by the Employer.

A question sometimes arises as to whether or not an implied order, in writing, to suspend part of the works constitutes an order under this clause. For example, the engineer instructs the contractor to proceed with some section of the works other than that shown on his programme because of a delay in the preparation of drawings. In such cases, it is submitted that the contractor should request an order to suspend the part of the works subject to delay and should, whether or not the order is issued, give notice of claim under clause 52(4).

In *John Jarvis Ltd* v. *Rockdale Housing Association Ltd* (1986) a main contractor under a JCT 80 building contract had problems with a nominated sub-contractor and wrote to the architect stating that as a result he was in difficulties and that it would be 'imprudent to proceed further . . . until it is known what is happening in connexion with the piles'. The architect's reply ('We agree with you that it would be imprudent to proceed any further with the work in connexion with the ground beams, and you should cease work on this element of the contract') was held to amount to an instruction to postpone execution of the work, and in the circumstances entitled the contractor to terminate the contract. The courts might well adopt a similarly realistic approach to problems under clause 40.

Provided that the contractor gives notice under clause 52(4) he is entitled to be paid the extra cost, with an addition thereto for profit, the sum being included in normal certificates. If the suspension is only in part occasioned by clause 40(1)(a), (b) or (c), the contractor gets the appropriate proportion of his extra cost and profit. Under the Fifth Edition there was no addition for profit.

Apart from the possible disruption of the contractor's programme, which may occasion delay and consequential costs, suspension of all or part of the works may result in additional costs due to inflation and in respect of slower working, for example where work such as excavation and bitumen macadam laying is thrown into the winter months.

See section 5.4 as to extension of time.

3.16.2 Minor Works Conditions

Clause 2.3(c) empowers the engineer to instruct suspension of the whole or part of the works under clause 2.6, which provides:

(1) The Engineer may order the suspension of the progress of the Works or any part thereof
  (a) for the proper execution of the work

(b)   for the safety of the Works or any part thereof

(c)   by reason of weather conditions

and in such event may issue such instructions as may in his opinion be necessary to protect and secure the Works during the period of suspension.

(2)   If permission to resume work is not given by the Engineer within a period of 60 days from the date of the written Order of Suspension then the Contractor may serve a written notice on the Engineer requiring permission to proceed with the Works within 14 days from the receipt of such notice. Subject to the Contractor not being in default under the Contract the Engineer shall grant such permission and if such permission is not granted the Contractor may by a further written notice served on the Engineer elect to treat the suspension where it affects a part of the Works as an omission under Clause 2.3(a) or where the whole of the Works is suspended as an abandonment of the Contract by the Employer.

Where suspension is so ordered, any extension of time is dealt with under clause 4.4(a) while the contractor's entitlement to extra costs falls to be determined under clause 6.

3.17   Possession of site

3.17.1   ICE Conditions

Clause 42 provides:

(1)   The Contract may prescribe

(a)   the extent of portions of the Site of which the Contractor is to be given possession from time to time

(b)   the order in which such portions of the Site shall be made available to the Contractor

(c)   the availability and the nature of the access which is to be provided by the Employer

(d)   the order in which the Works shall be constructed.

(2)   (a)   Subject to sub-clause (1) of this Clause the Employer shall give to the Contractor on the Works Commencement Date possession of so much of the Site and access thereto as may be required to enable the Contractor to commence and proceed with the construction of the Works.

(b)   Thereafter the Employer shall during the course of the Works give to the Contractor possession of such further portions of the Site as may be required in accordance with the programme which the Engineer has accepted under Clause 14 and such further access as is necessary to enable the Contractor to proceed with the construction of the Works with due despatch.

(3)   If the Contractor suffers delay and/or incurs additional cost from failure on the part of the Employer to give possession in accordance with the terms of this Clause the Engineer shall determine

(a)   any extension of time to which the Contractor is entitled under Clause 44 and

(b)   subject to Clause 52(4) the amount of any additional cost to which the Contractor may be entitled. Profit shall be added thereto in respect of any additional permanent or temporary work.

The Engineer shall notify the Contractor accordingly with a copy to the Employer.

(4)   The Contractor shall bear all costs and charges for any access required by him additional to those provided by the Employer. The Contractor shall also provide at his own cost any additional facilities outside the Site required by him for the purposes of the Works.

'The site' is defined in clause 1(1)(v) as meaning 'the lands and other places on under in or through which the Works are to be executed and any other lands or places provided by the Employer for the purposes of the Contract together with such other places as may be designated in the Contract or subsequently agreed by the Engineer as forming part of the Site'. It is important that the contract documents should clearly delineate the boundaries of the site and provide for partial possession of parts of the site if appropriate. The contractor's right to possession of the site is fundamental, and the employer's delay in giving sufficient possession is a breach which goes to the root of the contract. Under clause 42(2), the contractor's basic entitlement is to 'possession of so much of the site and access thereto as may be required to enable [him] to commence and proceed with the construction of the Works'.

The provisions of sub-clause 42(1) are important since the contract documents may dictate the order of the works and this in turn may be affected by facilities required for other contractors under clause 31(1).

Problems arise where the contract merely sets out the extent of the site without defining which parts of it will not be available throughout the construction period, and which parts of the site are not available at the commencement date. In such circumstances the contractor is entitled to the whole of the site – he may, for example, wish to use parts of it for storage, etc. and the contractor is entitled to claim an extension of time if delay is caused, as well as extra cost and profit thereon.

If the contract documents prescribe an order of working, then they must also make clear (if that is the case) that possession is to be given sequentially. It is not sufficient simply to state that the works are to be carried out in a particular order – it is necessary to 'prescribe the extent of portions of the site of which the Contractor is to be given possession from time to time'.

In a building contract case *(Rapid Building Group Ltd* v. *Ealing Family Housing Association Ltd* (1984)) where the possession clause provided that possession of the site should be given to the contractor on a specified date, part of the site was occupied by squatters. The employer was unable to give possession of the whole of the site on the due date and the squatters were not finally evicted for some 19 days. The Court of Appeal, upholding the trial judge, found that this was a breach of the contract by the employer that caused appreciable delay and entitled the contractor to damages. Under clause 42(1), delay and extra cost in such circumstances would be dealt with under the contract machinery.

Under the Fifth Edition there was no express reference to access to the site, although arguably such reference was implied; and the contractor was not entitled to profit on costs incurred through failure of the employer to give possession in accordance with the contract.

3.17.2 Minor Works Conditions

The position is much the same under the Minor Works form. Clause 4.1 provides:

> The starting date shall be the date specified in the Appendix or if no date is specified a date to be notified by the Engineer in writing being within a reasonable time and in any event within 28 days after the date of acceptance of the Tender. The Contractor shall begin the Works at or as soon as reasonably possible after the starting date.

Although this clause does not provide for the employer to give possession of the site in express terms, a term to that effect must necessarily be implied: *Freeman & Son* v. *Hensler* (1900). Clause 4.4(e) recognises 'failure by the Employer to give adequate access to the Works or possession of land required to perform the Works' as a ground for extension of time; it is also a ground for additional payment under clause 6.1.

In a Canadian case (The *Queen* v. *Walter Cabott Construction Ltd* (1975)) it was pointed out that in any construction contract 'it is fundamental to a building contract that work space be provided unimpeded by others'. For a variety of reasons access or possession may not be available as envisaged and hence both the ICE Conditions and the Minor Works Conditions seek to define who is to bear the financial consequences if possession or access is not given.

In *LRE Engineering Services Ltd* v. *Otto Simon Carves Ltd* (1981), which involved special conditions of contract requiring that 'access to and possession of the site ... be afforded to the contractor by [the employer] in proper time for the execution of the work', access was impeded by pickets over whom the employer had no control. This was held not to amount to a breach of contract by the employer. Clearly, the employer cannot be in breach of his obligation to give sufficient possession of the site to the contractor if he is wrongfully excluded from the site by third parties for whom the employer is not responsible in law and over whom he has no control. Access, in its ordinary and natural meaning, is physical access to the site. (This case is to be distinguished from *Rapid Building Group* above because the employer is deemed in law to be able to evict squatters but not lawful pickets.)

3.18   Rate of progress

Clause 46 of the ICE Conditions provides:

(1)   If for any reason which does not entitle the Contractor to an extension of time the rate of progress of the Works or any Section is at any time in the opinion of the Engineer too slow to ensure substantial completion by the time or extended time for completion prescribed by Clause 43 and 44 as appropriate or the revised time for completion agreed under sub-clause (3) of this Clause the Engineer shall notify the Contractor in writing and the Contractor shall thereupon take such steps as are necessary and to which the Engineer may consent to expedite the progress so as substantially to complete the Works or such Section by that prescribed time or extended time. The Contractor shall not be entitled to any additional payment for taking such steps.

(2)   If as a result of any notice given by the Engineer under sub-clause (1) of this Clause the Contractor shall seek the Engineer's permission to do any work on Site at night or on Sundays such permission shall not be unreasonably refused.

(3)   If the Contractor is requested by the Employer or the Engineer to

complete the Works or any Section within a revised time being less than the time or extended time for completion prescribed by Clauses 43 and 44 as appropriate and the Contractor agrees so to do then any special terms and conditions of payment shall be agreed between the Contractor and the Employer before any such action is taken.

Under clause 46, the engineer's power to order the contractor to expedite the works is dependent upon two provisos. Firstly, the rate of progress must, in the engineer's opinion, be too slow to ensure completion by the prescribed time or extended time for completion, as the case may be. Secondly, the reasons for the slow rate of progress must be of such a kind that they do not entitle the contractor to an extension of time under clause 44.

The engineer may be incorrect in his opinion that the rate of progress is inadequate to ensure completion by the due date. Progress is dependent upon a number of factors including, for example, 'learning time' – where a repetitive operation is involved – and weather. A contractor who challenges the need for expedition under clause 46 may later be able to demonstrate that the engineer's notification was unnecessary, in which case he could found a claim for reasonable payment in respect of the additional costs incurred.

The contractor may challenge the engineer's refusal to grant an extension of time under clause 44. Clause 46 refers to 'any reason which does not entitle the contractor to an extension of time' as though the entitlement were a simple matter of fact about which contention could not possibly arise. Clause 44 does, however, provide for two reviews by the engineer of his original decision, and there may be further reviews by the engineer, and by the arbitrator under clause 66.

It is arguable that, if under any of these reviews an extension of time is granted to the contractor, the clause 46 notice is invalidated where the extension is given on grounds which were in existence when the clause 46 notice was given. The question then arises as to whether or not the contractor may claim the costs incurred in complying with the engineer's notice to expedite progress.

Professor John Uff QC (*Keating on Building Contracts*, 5th edn, 1991, p875) notes that

'[al]though the matter is not free from doubt ... it is thought that an instruction under clause 46(1), which is retrospectively invalidated by an extension of time could take effect under clause 13(1) and give rise to a claim for compensation under clause 13(3). It would be necessary ... for the contractor to have applied for the extension of time in order to demonstrate that there was an entitlement at the date of the clause 46(1) notice.'

A contractor who considers a clause 46 notice to be invalid may simply refuse to comply with it. However, such action should not be taken lightly; it could incur the risk of the engineer's certifying under clause 63 that in his opinion the contractor has, despite previous warning, failed to proceed with the works with due diligence, or is persistently or fundamentally in breach of his obligations under the contract. Such a certificate under clause 63 could result in the expulsion of the contractor from the site, and at best give rise to protracted proceedings. If the contractor receives a clause 46 notice, it is thought that he should comply with it, record his dissent and, if necessary, invoke clause 66.

Sub-clause 46(3), which did not appear in the Fifth Edition, is arguably otiose. The contractor and the employer are, and always have been, free to negotiate and to agree terms for accelerating the work, whether or not their contract includes this sub-clause; and the involvement of the engineer in any such negotiations has similarly always been an option. If the sub-clause serves a purpose, it is to draw attention to that freedom and to clarify the fact that the engineer is not empowered to incur additional costs of acceleration unless specifically authorised by the employer to do so.

Apart from the absence of sub-clause 46(3) the position under the Fifth Edition was substantially the same. The opportunity has been taken in the Sixth Edition to improve the drafting of the clause, and to clarify the fact that the restrictions to working at night and on Sundays apply only to work on site.

There is no corresponding provision in the Minor Works Conditions.

3.19 Outstanding work and defects

3.19.1 ICE Conditions

Clause 49 states:

(1) The undertaking to be given under Clause 48(1) may after agreement between the Engineer and the Contractor specify a time or times within which the outstanding work shall be completed. If no such times are specified any outstanding work shall be completed as soon as practicable during the Defects Correction Period.

(2)   The Contractor shall deliver up to the Employer the Works and each Section and part thereof at or as soon as practicable after the expiry of the relevant Defects Correction Period in the condition required by the Contract (fair wear and tear excepted) to the satisfaction of the Engineer. To this end the Contractor shall as soon as practicable execute all work of repair amendment reconstruction rectification and making good of defects of whatever nature as may be required of him in writing by the Engineer during the relevant Defects Correction Period or within 14 days after its expiry as a result of an inspection made by or on behalf of the Engineer prior to its expiry.

(3)   All work required under sub-clause (2) of this Clause shall be carried out by the Contractor at his own expense if in the Engineer's opinion it is necessary due to the use of materials or workmanship not in accordance with the Contract or to neglect or failure by the Contractor to comply with any of his obligations under the Contract. In any other event the value of such work shall be ascertained and paid for as if it were additional work.

(4)   If the Contractor fails to do any such work as aforesaid the Employer shall be entitled to carry out such work by his own work-people or by other contractors and if such work is work which the Contractor should have carried out at his own expense the Employer shall be entitled to recover the cost thereof from the Contractor and may deduct the same from any monies that are or may become due to the Contractor.'

The opportunity has been taken in the Sixth Edition to replace the inaccurate and misleading term 'maintenance period' used in earlier editions by the more precise 'defects correction period': a term that first appeared in the Minor Works Conditions. Provision has also been made for agreement between the engineer and the contractor as to the programme for carrying out remedial works; and the verbose and complex provisions of the Fifth Edition relating to temporary reinstatement of roads have been omitted.

Defects in the works have been categorised as those caused by 'fair wear and tear' and those requiring 'repair, amendment, reconstruction, rectification and making good'. Within the second category defects are sub-divided into

(a)   those 'due to the use of materials or workmanship not in accordance with the Contract or to neglect or failure by the Contractor to comply with any of his obligations under the Contract', and

(b)   those resulting from any other event.

The contractor is not bound to make good defects caused by wear and tear, but terms for so doing could of course be agreed between the parties. Benefit to both parties could arise from such an agreement. Where, for example, crash barriers on a motorway are damaged by third-party traffic during the defects correction period, the contractor, while not obliged to do so, might offer an economic price for the repairs, having all necessary facilities on site.

The programme for repair works relates to clause 48(1), and could be a factor influencing the issue of the certificate of substantial completion. Apart from any such agreement, the contractor's obligations are to execute repairs 'as soon as practicable' and 'to deliver up to the Employer the Works and each Section and part thereof at or as soon as practicable after the expiry of the relevant Defects Correction Period'.

Defects that appear during the defects correction period and which are notified in writing by the engineer during that period or within 14 days of its expiry must be remedied by the contractor. If those defects fall within category (a) above the cost of putting them right is borne by the contractor. If the defects result from any other cause the contractor is entitled to payment if he puts them right, and they fall to be valued under clause 52(1)(b) as additional work ordered during the defects correction period.

The burden of proving that the contractor has failed to comply with his contractual obligations lies on the employer. Thus, for example, where cracks appear in a concrete structure and the contractor alleges that they were caused by inadequate design, the engineer would need to have evidence that those cracks were caused by some breach by the contractor in order to resist a claim for the cost of remedial works.

3.19.2   Minor Works Conditions

The Minor Works Conditions refer also to the 'defects correction period'. Clause 5 of those conditions provides as follows:

> 5.1 'Defects Correction Period' means the period stated in the Appendix which period shall run from the date certified as practical completion of the whole of the Works or the last period thereof.
> 5.2 If any defects appear in the Works during the Defects Correction Period which are due to the use of materials or workmanship not in accordance with the Contract the Engineer shall give written notice thereof and the Contractor shall make good the same at his own cost.

5.3   If any such defects are not corrected within a reasonable time by the Contractor the Employer may after giving 14 days written notice to the Contractor employ others to correct the same and the cost thereof shall be payable by the Contractor to the Employer.

5.4   Upon the expiry of the Defects Correction Period and when any outstanding work notified to the Contractor under Clause 5.2 has been made good the Engineer shall upon the written request of the Contractor certify the date on which the Contractor completed his obligations under the Contract to the Engineer's satisfaction.

5.5   Nothing in Clause 5 shall affect the rights of either party in respect of defects appearing after the Defects Correction Period.

The defects correction period in the case of minor works is normally six months, and the *Notes for Guidance* suggest that it should never exceed 12 months. During it, the contractor's liability (see clause 5.2) is limited to making good at his own cost any defects which appear during the period which are due to materials or workmanship not in accordance with the contract. However clause 3.2 clearly envisages that the contractor can achieve completion despite the fact that items of work are outstanding, since it refers to the contractor's responsibility 'for the care of any outstanding work which he has undertaken to finish during the Defects Correction Period until such outstanding work is complete'.

This is reinforced by the provisions of clause 4.7, which is in the following terms:

4.7   The Contractor shall rectify any defects and complete any outstanding items in the Works or any part thereof which reach practical completion promptly thereafter or in such manner and/or time as may be agreed or otherwise accepted by the Engineer. The Contractor shall maintain any parts which reach practical completion in the condition required by the Contract until practical completion of the whole of the Works fair wear and tear excepted.

Once again, the exception in respect of 'fair wear and tear' should be noted.

3.20   *Contractor to search for defects etc.*

Clause 50 of the ICE Conditions states:

The Contractor shall if required by the Engineer in writing carry out such searches tests or trials as may be necessary to determine the cause of any defect imperfection or fault under the directions of the Engineer. Unless such defect imperfection or fault shall be one for which the Contractor is liable under the Contract the cost of the work carried out by the Contractor as aforesaid shall be borne by the Employer. But if such defect imperfection or fault shall be one for which the Contractor is liable the cost of the work carried out as aforesaid shall be borne by the Contractor and he shall in such case repair rectify and make good such defect imperfection or fault at his own expense in accordance with Clause 49.

This clause recognises the likelihood that the cause of a defect, imperfection or fault may not be immediately apparent. It may be necessary to carry out searches, tests or trials; and the rules for determining responsibility for the costs incurred correspond with those applicable to repair works under clause 49. The burden of proof once again lies upon the employer. It is for him or the engineer to show that the defects were caused by some breach by the contractor if a claim for cost under the clause is to be resisted.

The engineer's powers under clause 50 are not only exercisable during the defects correction period, and it has been suggested (Abrahamson, *Engineering Law and the ICE Contracts,* 4th edn, p168) that the engineer's requirements under clause 50 could amount to an engineer's instruction entitling the contractor to payment for delay and costs other than the 'cost of the work' specifically referred to in the clause. This seems to us to be correct.

Clause 50 is unchanged from that clause of the Fifth Edition.

There is no exactly comparable clause in the Minor Works Conditions, but the engineer is empowered to order the 'carrying out of any test or investigation' under clause 2.3(b), and if those are ordered and the results are in favour of the contractor then he is entitled to an extension of time under clause 4.4(b) and additional payment under clause 6.1.

3.21 Provisional and prime cost sums and nominated sub-contractors

Clause 58 of the ICE Conditions states:

(1) In respect of every Provisional Sum the Engineer may order either or both of the following.
  (a) Work to be executed or goods materials or services to be supplied by the Contractor the value thereof being determined in accordance with Clause 52 and included in the Contract Price.

(b)   Work to be executed or goods materials or services to be supplied by a Nominated Sub-contractor in accordance with Clause 59.

(2)   In respect of every Prime Cost Item the Engineer may order either or both of the following.

(a)   Subject to Clause 59 that the Contractor employ a sub-contractor nominated by the Engineer for the execution of any work or the supply of any goods materials or services included therein.

(b)   With the consent of the Contractor that the Contractor himself execute any such work or supply any such goods materials or services in which event the Contractor shall be paid in accordance with the terms of a quotation submitted by him and accepted by the Engineer or in the absence thereof the value shall be determined in accordance with Clause 52 and included in the Contract Price.

(3)   If in connection with any Provisional Sum or Prime Cost Item the services to be provided include any matter of design or specification of any part of the Permanent Works or of any equipment or plant to be incorporated therein such requirement shall be expressly stated in the Contract and shall be included in any Nominated Sub-contract. The obligation of the Contractor in respect thereof shall be only that which has been expressly stated in accordance with this sub-clause.

In the first edition of this book the authors described the provisions for nominated sub-contracting contained in clauses 58, 59A, 59B and 59C of the Fifth Edition as being the most complex, and arguably the least satisfactory provisions of the ICE Conditions. They recognised, however, that these defects arose from the fundamentally unsatisfactory nature of nomination: and that those who had drafted the relevant clauses had done so (in full cognisance of the fundamental defects) because of the greater danger that, if no provision were made for nomination, even less satisfactory clauses might be drafted by employers or engineers on an *ad hoc* basis. The basic objections to nomination apply equally to the provisions of the Sixth Edition; but further safeguards have been introduced to protect a contractor who has valid reason to object to a nominee.

Clauses 58 and 59 need to be read together, and they attempt to solve the many legal and practical difficulties created by the system of nominated sub-contracting. A basic practical problem is that of imposing on the main contractor a sub-contractor not of his choosing, unknown to him at the time of making the main contract, and in whom he may have little, if any, confidence. The draftsmen have attempted to protect the main contractor against defaults by the nominee by including procedures at each stage of the nomination procedure designed to provide such protection.

As a result, under clauses 58 and 59 the employer may become financially responsible for losses resulting from the nominated sub-contractor's default, to the extent that the main contractor is unable to recover his losses from the defaulting sub-contractor. Furthermore, under provisions introduced in the Sixth Edition, the employer may be unable to require the main contractor to enter into a nominated sub-contract, and may have to make other arrangements. Ideally, this should have the desirable effect of discouraging the use of nomination where it can be avoided, as it can be in many instances. For example, the main contract documents may specify the names of one or more approved sub-contractors whom the main contractor may employ as his domestic sub-contractors or alternatively the employer may enter into direct contracts with specialists of his own choice.

Another important problem is that of design liability. In many instances the main purpose of nomination is to enlist the nominee's specialist design knowledge. Unlike its predecessor the Sixth Edition recognises, in sub-clauses 7(6) and 8(2), the possibility that the contractor's obligations may include design of part of the permanent works – consequently special provision to cover the design element in a nominated sub-contract is no longer needed. Sub-clause 58(3) does however require an express statement of design services included in a nominated sub-contract.

Definitions of 'provisional sum', 'prime cost (PC) item', and 'nominated sub-contractors', previously given in sub-clauses 58(1), (2) and (5) of the Fifth Edition, now appear, more logically, among the definitions in clause 1. The meaning of 'nominated sub-contractor', which was defined in sub-clause 58(5) of the Fifth Edition, is now defined in sub-clause 1(1)(m) of the Sixth Edition, and establishes that, in contrast to usual construction industry semantics, the term applies equally to a case where the sub-contract is for supply only.

Clause 58(1) clarifies the engineer's powers in relation to provisional sums. He may either require the main contractor to execute the work or supply the goods, materials or services or else appoint a nominated sub-contractor under the provisions of clause 59. He is, of course, also empowered to omit such work etc. The engineer's powers in relation to provisional sums are defined in sub-clause 58(1), and make it clear that he may decide whether or not to expend such sums at all. In contrast, a prime cost item will be used for the execution of work or for the supply of goods, materials and services for the works – hence the omission of any such work can be ordered only as a variation, which then falls to be valued under clause 52.

If, following a successful objection to a nominee by the contractor, the engineer orders a variation omitting a PC item under clause 59(2)(c), the contractor is entitled to have included in the contract price 'such sum (if any) in respect of [his] charges and profit being a percentage of the estimated value of such omission as would have been payable had there been no such omission'.

Sub-clauses 58(2)(d) and 58(2)(e) appear to call in question the reason why provision for nomination was made in the main contract. However it should be recognised that the main contract documents were prepared without knowledge of the capabilities of the main contractor in relation to any specialist work included. For example, where the main contractor is found to have a specialist piling division it would be absurd to exclude that division from consideration as nominee. Of course the contractor's consent is necessary before the engineer can order him to do the PC item work, which may restrict the operation of the provision in practice.

The Minor Works form does not provide for provisional items or PC sums.

3.22 *Nominated sub-contractors – clause 59*

Clause 59 of the ICE Conditions states:

(1) The Contractor shall not be under any obligation to enter into a sub-contract with any Nominated Sub-contractor against whom the Contractor may raise reasonable objection or who declines to enter into a sub-contract with the Contractor containing provisions

(a) that in respect of the work goods materials or services the subject of the sub-contract the Nominated Sub-contractor will undertake towards the Contractor such obligations and liabilities as will enable the Contractor to discharge his own obligations and liabilities towards the Employer under the terms of the Contract

(b) that the Nominated Sub-contractor will save harmless and indemnify the Contractor against all claims demands and proceedings damages costs charges and expenses whatsoever arising out of or in connection with any failure by the Nominated Sub-contractor to perform such obligations or fulfil such liabilities

(c) that the Nominated Sub-contractor will save harmless and indemnify the Contractor from and against any negligence by the Nominated Sub-contractor his agents workmen and servants and against any misuse by him or them of any Contractor's Equipment or Temporary Works provided by the Contractor for the purposes of the Contract and for all claims as aforesaid

(d)   that the Nominated Sub-contractor will provide the Contractor with security for the proper performance of the sub-contract and

(e)   equivalent to those contained in Clause 63.

(2)   If pursuant to sub-clause (1) of this Clause the Contractor declines to enter into a sub-contract with a sub-contractor nominated by the Engineer or if during the course of the Nominated Sub-contract the Contractor shall validly terminate the employment of the Nominated Sub-contractor as a result of his default the Engineer shall

(a)   nominate an alternative sub-contractor in which case sub-clause (1) of this Clause shall apply or

(b)   by order under Clause 51 vary the Works or the work goods materials or services in question or

(c)   by order under Clause 51 omit any or any part of such works goods materials or services so that they may be provided by workmen contractors or suppliers employed by the Employer either

(i)   concurrently with the Works (in which case Clause 31 shall apply) or

(ii)   at some other date

and in either case there shall nevertheless be included in the Contract Price such sum (if any) in respect of the Contractor's charges and profit being a percentage of the estimated value of such omission as would have been payable had there been no such omission and the value thereof had been that estimated in the Bill of Quantities or inserted in the Appendix to the Form of Tender as the case may be or

(d)   instruct the Contractor to secure a sub-contractor of his own choice and to submit a quotation for the work goods materials or services in question to be so performed or provided for the Engineer's consideration and action or

(e)   invite the Contractor himself to execute or supply the work goods materials or services in question under Clause 58(1)(a) or Clause 58(2)(b) or on a daywork basis as the case may be.

(3)   Except as otherwise provided in Clause 58(3) the Contractor shall be as responsible for the work executed or goods materials or services supplied by a Nominated Sub-contractor employed by him as if he had himself executed such work or supplied such goods materials or services.

(4)   (a)   If any event arises which in the opinion of the Contractor justifies the exercise of his right under any Forfeiture Clause to terminate the sub-contract or to treat the sub-contract as repudiated

by the Nominated Sub-contractor he shall at once notify the Engineer in writing.

(b) With the consent in writing of the Engineer the Contractor may give notice to the Nominated Sub-contractor expelling him from the Sub-contract works pursuant to any Forfeiture Clause or rescinding the Sub-contract as the case may be. If however the Engineer's consent is withheld the Contractor shall be entitled to appropriate instructions under Clause 13.

(c) In the event that the Nominated Sub-contractor is expelled from the Sub-contract works, the Engineer shall at once take such action as is required under sub-clause (2) of this Clause.

(d) Having with the Engineer's consent terminated the Nominated Sub-contract the Contractor shall take all necessary steps and proceedings as are available to him to recover all additional expenses that are incurred from the Sub-contractor or under the security provided pursuant to sub-clause (1)(d) of this Clause. Such expenses shall include any additional expenses incurred by the Employer as a result of the termination.

(e) If and to the extent that the Contractor fails to recover all his reasonable expenses of completing the Sub-contract works and all his proper additional expenses arising from the termination the Employer will reimburse the Contractor his unrecovered expenses.

(f) The Engineer shall take any delay to the completion of the Works consequent upon the Nominated Sub-contractor's default into account in determining any extension of time to which the Contractor is entitled under Clause 44.

(5) For all work executed or goods materials or services supplied by Nominated Sub-contractors there shall be included in the Contract Price

(a) the actual price paid or due to be paid by the Contractor in accordance with the terms of the Sub-contract (unless and to the extent that any such payment is the result of a default of the Contractor) net of all trade and other discounts rebates and allowances other than any discount obtainable by the Contractor for prompt payment

(b) the sum (if any) provided in the Bill of Quantities for labours in connection therewith and

(c) in respect of all other charges and profit a sum being a percentage of the actual price paid or due to be paid calculated (where provision has been made in the Bill of Quantities for a rate to be set against the relevant item of prime cost) at the rate inserted by the Contractor against that item or (where no such provision has been made) at the rate inserted by the Contractor in the Appendix to the Form of Tender as the percentage for adjustment of sums set against Prime Cost Items.

(6)   The Contractor shall when required by the Engineer produce all quotations invoices vouchers sub-contract documents accounts and receipts in connection with expenditure in respect of work carried out by all Nominated Sub-contractors.

(7)   Before issuing any certificate under Clause 60 the Engineer shall be entitled to demand from the Contractor reasonable proof that all sums (less retentions provided for in the Sub-contract) included in previous certificates in respect of the work executed or goods or materials or services supplied by Nominated Sub-contractors have been paid to the Nominated Sub-contractors or discharged by the Contractor in default whereof unless the Contractor shall

    (a)   give details to the Engineer in writing of any reasonable cause he may have for withholding or refusing to make such payment and

    (b)   produce to the Engineer reasonable proof that he has so informed such Nominated Sub-contractor in writing

the Employer shall be entitled to pay to such Nominated Sub-contractor direct upon the certification of the Engineer all payments (less retentions provided for in the Sub-contract) which the Contractor has failed to make to such Nominated Sub-Contractor and to deduct by way of set-off the amount so paid by the Employer from any sums due or which become due from the Employer to the Contractor. Provided always that where the Engineer has certified and the Employer has made direct payment to the Nominated Sub-contractor the Engineer shall in issuing any further certificate in favour of the Contractor deduct from the amount thereof the amount so paid but shall not withhold or delay the issue of the certificate itself when due to be issued under the terms of the Contract.

Clause 59, which is a concise amalgam of clauses 59A, 59B and 59C of the Fifth Edition, is to be welcomed if for no other reason than that it reduces the appalling verbosity of the earlier provisions. Besides doing so, however, it also makes important changes in the principles governing nomination.

3.22.1   Objections to nomination

As in clause 59A of the Fifth Edition, clause 59 provides a right of reasonable objection by the main contractor to the engineer's nominee, and the main contractor is not bound to sub-contract with a nominee who refuses to undertake the main contractor's obligations in respect of the sub-contracted work. Although no guidance is given as to what is a 'reasonable objection' it is suggested that grounds for such objection would include incompatibility of the nominee's programme with the contractor's programme, financial instability, the nominee's lack of ability or specialist skills and so on.

Where the contractor makes a reasonable objection the engineer has five options under sub-clause 59(2):

- To nominate another sub-contractor. The main contractor would then retain his right of reasonable objection.
- To vary the works or work goods materials or services under clause 51.
- To omit the works etc. and arrange for them to be performed by others employed directly by the employer, either concurrently with the main contract works or at some other date. The contractor would then be entitled to his charges and profit on the omitted work etc.
- To instruct the main contractor to obtain a quotation from a sub-contractor of his own choice and to submit such quotation for the engineer's approval and action. In such a case it is reasonable to infer that the main contractor would accept full responsibility for the sub-contractor's performance as if he were a domestic sub-contractor.
- To arrange for the contractor to carry out the work etc. In the case of a provisional sum the contractor is bound to do the work, but not in the case of a PC item when his consent is necessary: see clause 58.

A provision of clause 59A of the Fifth Edition, under which the engineer had power to order the contractor to enter into a sub-contract with a nominee against whom he had a valid objection, on the employer's accepting responsibility for the risks giving rise to the objections has (wisely in our opinion) been omitted.

Clause 59(3) makes the main contractor responsible for the quality of the sub-contractor's work as if the work or supply had been by himself or his domestic sub-contractor, except as otherwise provided in clause 58(3). The exception refers to the contractor's obligations in respect of any design or specification element included in the sub-contract, which obligations cover only those matters that have been expressly stated in accordance with sub-clause 58(3). This provision remains substantially unaltered from that covered by sub-clause 59A(4) of the Fifth Edition.

3.22.2 Forfeiture of nominated sub-contract

The situation that arises where a default by the sub-contractor is such as to warrant his expulsion from the works is covered by sub-clause 59(4). The contractor is required in such circumstances at once to give written notice to the engineer of his intention to terminate the sub-contract or to treat it as repudiated; and where the engineer concurs with such action the sub-contractor may be expelled. Thereafter the contractor must, under sub-clause 59(4)(d), take all necessary steps and proceedings in order to recover his losses from the sub-contractor – but any losses incurred by the contractor in completing the sub-contract works that he is unable to recover from the sub-contractor (or other expenses arising from the termination) are recoverable from the employer.

Where the engineer withholds his consent to termination of the sub-contract he must give instructions pursuant to clause 13, which provides *inter alia* for extension of time in the event that delay is caused and for payment to the contractor of costs incurred plus profit thereon.

The principal objective of this clause is to ensure that the employer's right under clause 63 to terminate the contract in the event of the insolvency of the main contractor or his abandonment of the contract or other serious breach is reflected in a corresponding right of the main contractor against the nominated sub-contractor in the nominated sub-contract. The clause also sets out the procedure to be followed and defines the payments to which the main contractor is entitled in the event of valid forfeiture of the sub-contract. Clause 16 of the Civil Engineering Contractors Association (CECA) (formerly the Federation of Civil Engineering Contractors) Form of Sub-contract is the corresponding provision.

Forfeiture and its consequences are adequately discussed in the legal textbooks. This commentary is confined to the 'claims' aspects of this clause which, while less complex than its predecessor in the Fifth Edition, remains an extremely complicated clause.

Delay and extra cost consequent on the forfeiture of a nominated sub-contract are covered by clause 59(4)(d). It is important that the forfeiture procedure should have been followed: that is, that the engineer's consent to the forfeiture shall have been obtained. In the absence of such consent the main contractor is unable to rely on the protection given him by sub-clauses 59(4)(d), (e) and (f); and he will of course be liable in damages to the sub-contractor. The perils of forfeiture are spelled out in

Powell-Smith and Sims, *Determination and Suspension of Construction Contracts*, 1985 pp124–33.

Clause 59(5) covers the basis of payment in respect of work etc. executed by a nominated sub-contractor and is unchanged from sub-clause 59A(5) of the Fifth Edition. It impliedly confirms that the percentage addition to the PC sum, whether included in the bill of quantities as a separate item or in the Appendix to the Form of Tender as a general provision, covers only administrative work and profit.

3.22.3 Payments to nominated sub-contractors

Clause 59(7) deals with payments by the contractor to the nominated sub-contractor and repeats the provisions of clause 59C of the Fifth Edition under which the engineer is entitled to certify direct payments by the employer to a nominated sub-contractor in certain circumstances. One of the unfortunate effects of that provision is that it tends to aggravate the antagonism that so often exists between the main contractor and the nominee. The nominated sub-contractor knows that he was appointed by the engineer, that the main contractor was possibly a reluctant party, and that it is to the engineer that he must look for future contracts. In commercial practice, therefore, he looks for the engineer's approval rather than that of the main contractor and antagonism often arises between the main contractor and the nominated sub-contractor from this cause.

A common cause of contention about payments to a nominated sub-contractor is an allegation by the main contractor that the nominee has delayed his work or has executed defective work or has in some way caused the main contractor to incur additional liabilities or costs.

The important safeguard for the contractor is that direct payment may be made only in strict accordance with the provisions of clause 59(7) and against the certificate of the engineer which must take into account 'any reasonable cause' that the contractor may have for withholding or refusing to pay the nominated sub-contractor. It is thought that delay by a nominated sub-contractor would constitute a reasonable cause.

The CECA Form of Sub-contract, clause 15, deals with payments by the contractor to the sub-contractor. There is no substantial change to that clause in comparison with the 1984 and 1991 editions of the FCEC Form other than a change of terminology in that 'maintenance' has been renamed 'defects correction' in order to maintain consistency with the Sixth Edition. In *NEI Thomson Ltd* v. *Wimpey Construction Ltd* (1987) the Court of Appeal held that the main contractor's ordinary right to rely on set-off at common law in respect of some dispute – in that case a crossclaim for damages for delay – was not excluded by the wording of clause 15(3) of the FCEC Form of Sub-contract.

3.23 Interest on overdue payments

3.23.1 ICE Conditions

Clause 60(7) of the ICE Conditions states:

(7) In the event of
(a) failure by the Engineer to certify or the Employer to make payment in accordance with sub-clauses (2) (4) or (6) of this Clause or
(b) any finding of an arbitrator to such effect
the Employer shall pay to the Contractor interest compounded monthly for each day on which any payment is overdue or which should have been certified and paid at a rate equivalent to 2% per annum above the base lending rate of the bank specified in the Appendix to the Form of Tender. If in an arbitration pursuant to Clause 66 the arbitrator holds that any sum or additional sum should have been certified by a particular date in accordance with the aforementioned sub-clauses but was not so certified this shall be regarded for the purpose of this sub-clause as a failure to certify such sum or additional sum. Such sum or additional sum shall be regarded as overdue for payment 28 days after the date by which the arbitrator holds that the Engineer should have certified the sum or if no such date is identified by the arbitrator shall be regarded as overdue for payment from the date of the Certificate of Substantial Completion for the whole of the Works.

Clause 60(7) of the Sixth Edition is a substantial revision of the corresponding provision (clause 60(6)) of the Fifth Edition which, in material part, provided that 'in the event of failure by the Engineer to certify or the Employer to make payment in accordance with [the relevant sub-clauses] the Employer shall pay to the Contractor interest upon any payment overdue thereunder'. The interpretation of this provision was a matter of controversy and there was conflicting English and Scottish case law.

Fifth Edition provisions

The dispute was two-fold. Firstly, clause 60(2), which remains unaltered in the Sixth Edition, required the engineer to certify 'the amount which in the opinion of the Engineer … is due to the contractor'. It was thus arguable that, where a claim had been rejected by the engineer because in his opinion it was invalid, a later decision of the engineer or of an arbitrator under clause 66 that the claim was valid and payable did not mean that the engineer's original rejection of the claim amounted to a 'failure to certify'.

Max Abrahamson in his book *Engineering Law and the ICE Conditions*, 6th edn, 1979, p268, commented on the provision:

> 'There is an unfortunate drafting error. Sub-cls (2) and (3) require the engineer to certify the amount in his "opinion" due to the contractor, so that it might be argued that there is a failure "… to certify … in accordance with sub-clauses (2) and (3)" so as to attract interest only where the engineer acts in bad faith by certifying less than is due in his opinion, not merely when his opinion is later found to be wrong.'

Secondly, it was not clear in these circumstances whether, if the engineer failed to certify (and therefore the employer to pay) in full the amount of the sum claimed and to which the contractor could subsequently prove his entitlement, the employer was required to pay *compound* interest.

It was also arguable, under the above construction of clause 60(6) of the Fifth Edition, that the parties, having agreed to terms under which there was no contractual right to interest on the overdue payment, had impliedly agreed that the arbitrator ought not to exercise his discretion to award interest under the Arbitration Act, frustrating that agreement. The consequences of such a construction were that the employer and his engineer knew that it was very much in the employer's interests that payment be withheld where doubt existed as to the validity of claims arising from the contract. Interest on any sum later found to be due might not become payable at all; and even if the engineer or the arbitrator later decided that interest was payable, it was likely to be simple rather than compound – at least for the period following issue of the final certificate.

In two Scottish cases (*Nash Dredging Ltd* v. *Kestrel Marine Ltd* (1968) and *Hall and Tawse Ltd* v. *Strathclyde Regional Council* (1990)) it had been decided firstly that only simple interest was payable and further that the word 'failure' meant 'no more than that the engineer had not done something which he ought to have done in the performance of his functions under sub-clause (2). There would not be a failure on the part of the engineer to certify merely because the sum certified turned out to be less than the sum which the court or an arbiter (arbitrator) considered was due'.

Neither of these decisions was cited to the Official Referee in *Morgan Grenfell (Local Authority Finance) Ltd* v. *Seven Seas Dredging Ltd (No 2)* (1990), where the issue was argued by the respondent on the basis that 'failure to certify' referred only to an engineer who did not act *bona fide*. In that case, the arbitrator had found firstly that a failure by the engineer to certify sums that he (the arbitrator) later found to be payable constituted a 'failure to certify', and secondly that since the contractor was entitled to monthly payments in respect of interest on overdue payments, the contractor was entitled to have unpaid interest compounded monthly with the principal sum. That decision was expressly upheld by Judge Newey QC.

Later Mr Justice Hobhouse refused to accept the decision of Judge Newey in *Morgan Grenfell* and in *Secretary of State for Transport* v. *Birse-Farr Joint Venture* (1993), held, *inter alia*, firstly that:

'The words "failure by the engineer to certify" refer to, and refer only to, some failure of the engineer which can be identified as a failure by the engineer to respect and give effect to the provisions of the contract. Those words do not refer to under-certification which does not involve any contractual error or misconduct of the engineer.'

and secondly that:

'What clause 60(6) makes provision for is the payment of contractual interest not compound interest.'

However, as Mr Justice Hobhouse observed:

'that does not leave the contractor in a position where he cannot compound the interest. The interest is payable as a contractual sum due under the contract. If, for example, a previous certificate has not been paid by the employer, ... simple interest becomes payable upon the unpaid sum. There is no reason why the contractor should not be entitled to include a claim for such interest in a subsequent monthly statement ... [and then] the engineer is obliged to consider whether or not there was a liability to interest on a sum which the employer has failed to pay and if so to include the interest in the sum which he then certifies. If the employer continues to fail to pay, then the same procedure can be followed in the following months and the effect will be that the contractual interest is capitalized and recalculated in each succeeding certificate (up to the final certificate) until liability is discharged by payment.'

He went on to say that the situation was the same if the engineer wrongfully refuses to certify payment in disregard of the provisions of the contract, but 'since the power to compound the interest is related to the system of monthly payments and monthly certificates it would appear to follow that the right to compound does not continue after the issue of the final certificate'.

Sixth Edition clarification

Clause 60(7) of the Sixth Edition clarifies the position, in that wherever there is a 'failure by the Engineer to certify or the Employer to make payment ... or ... any finding of an arbitrator to such effect' interest is payable and that interest is to be compounded at monthly intervals. Furthermore the phrase 'failure to certify' is defined in the sub-clause to include cases where the arbitrator holds that 'any sum or additional sum should have been certified by a particular date in accordance with the aforementioned sub-clauses but was not so certified' – making it clear that Mr Justice Hobhouse's construction of the words used in the earlier form was not that intended by its drafters.

Where for any reason the contractual right to interest on late payments is not applicable, a fall-back provision is now available in arbitration under section 49 of the Arbitration Act 1996, under which (subject to the right of the parties to agree otherwise) the arbitrator is empowered to award '... simple or compound interest from such dates, at such rates and with such rests as [he] considers meets the justice of the case'. Furthermore that power is available on any sum awarded by the arbitrator in respect of any period up to the date of the award, and on any sum forming any part of a claim in the arbitration but paid before the date of the award, in respect of any period up to the date of payment. In addition the arbitrator may award simple or compound interest on sums awarded from the date of the award until payment is made.

The potential injustice resulting from the words used in sub-clause 60(6) of the Fifth Edition has been corrected in the corresponding sub-clause, namely 60(7), of the Sixth Edition – and the power to award such interest as will provide reasonable recompense to the contractor for having been kept out of his money is reinforced by section 49 of the Arbitration Act 1996, under which the arbitrator has a discretionary power to award simple *or compound* interest. Contractors should however be wary, at the time of making contracts, of attempts to nullify these improvements – either by the adoption of the Fifth Edition as the form of contract or by modifications to clause 60 of the Sixth Edition, and possibly the deletion of clause 66.

3.23.2 Minor Works Conditions

Clause 7.8 of the Minor Works form states:

> In the event of failure by the Engineer to certify or the Employer to make payment in accordance with the Contract or any finding of an arbitrator to such effect the Employer shall pay to the Contractor interest compounded monthly on the amount which should have been certified or paid on a daily basis at a rate equivalent to 2% per annum above the base lending rate of the bank specified in the Appendix.

This is a simpler version of the equivalent clause in the Sixth Edition.

CHAPTER FOUR
VARIATIONS

4.1 Duty/power of engineer to order variations

Clause 51 of the Sixth Edition reads:

(1) The Engineer
  (a) shall order any variation to any part of the Works that is in his opinion necessary for the completion of the Works and
  (b) may order any variation that for any other reason shall in his opinion be desirable for the completion and/or improved functioning of the Works.
Such variations may include additions omissions substitutions alterations, changes in quality form character kind position dimension level or line and changes in any specified sequence method or timing of construction required by the Contract and may be ordered during the Defects Correction Period.
(2) All variations shall be ordered in writing but the provisions of Clause 2(6) in respect of oral instructions shall apply.
(3) No variation ordered in accordance with sub-clauses (1) and (2) of this Clause shall in any way vitiate or invalidate the Contract but the value (if any) of all such variations shall be taken into account in ascertaining the amount of the Contract Price except to the extent that such variation is necessitated by the Contractor's default.
(4) No order in writing shall be required for increase or decrease in the quantity of any work where such increase or decrease is not the result of an order given under this Clause but is the result of the quantities exceeding or being less than those stated in the Bill of Quantities.

The basic purpose of clause 51 is to empower the engineer unilaterally to order variations of the works. In the absence of such a power, the works could only be varied by agreement between the contracting parties. Without a variation clause an engineer who found, for example, that his provision for foundations was inadequate could only seek to vary his design by persuading the contractor to allow the change. Knowing that without the variation the engineer's design would fail, the contractor would be able to demand an exorbitant price for it or to refuse to allow it at all.

Apart from variations that are *necessary*, and in respect of which the clause imposes a *duty* upon the engineer to order them, there is another category of variations; these are variations that are *desirable*, and which the engineer is *empowered* to order, although he is not under a duty to do so.

Within the first category are those variations that are necessary in order to correct errors in the design (which in this context includes drawings and specification) and those necessary to deal with unforeseen conditions. The latter might include unknown subsoil conditions, physical or artificial obstructions, or perhaps even the destruction by fire of the factory from which some specified material was to have been obtained.

Desirable variations might include those needed to take advantage of some improved technique intended to be used in the structure being constructed under the contract or to provide for an expanding or reducing demand for the product, to use some improved construction technique or material, to economise on foundations where subsoil conditions are unexpectedly favourable, or simply to accommodate the employer's changed wishes.

The duty/power to vary is not, however, completely unfettered. Variations in the first category must be those 'necessary for the completion of the Works'. Those in the second category must be 'desirable for the completion and/or improved functioning of the Works'. Hence, for example, it would not be within the power or duty of an engineer who finds an exceptionally low price in a contract to order the contractor to carry out work at that price at some other location unconnected with the works as defined in the contract because this is not 'necessary for the completion of the Works'. Again, clause 51 could not be used for a variation that alters the nature of the works – 'the Works' must remain substantially as defined at the time of making the contract.

Under clause 13, the duty/power to give directions, which term must include variations, is vested only in the engineer and, to the extent that he has delegated power, in the engineer's representative. It follows that neither the employer nor an RE (unless he has delegated authority) is empowered to order variations.

Within these limitations, the scope for variation is wide indeed. It includes additions, omissions, substitutions, alterations, changes in quality, form, character, kind, position, dimension, level or line, and changes in the specified sequence, method or timing of construction (if any). With the exception of a variation which alters the *nature* of the works, this definition must cover almost any change likely to be required during the execution of civil engineering works. But the power cannot be used to alter the fundamental nature of the works defined in the contract.

An express power to order variation during the defects correction period is now included – clause 51(1). No such power was contained in the Fifth Edition.

Variations agreed by the engineer at the request of the contractor are in a separate category. For example, where the contractor wishes to vary the works in some way in order to make use of plant or materials that he owns, such as a particular type of piling rig, the engineer's agreement to the variation is usually formalised by the issue of a variation order. In such instances, it is important that the variation order should clarify the reason why it was issued and should, where appropriate, confirm that it will not entitle the contractor to additional payment.

This will, however, not automatically be the case. The contractor may, for example, suggest the adoption of his own piling system in preference to that defined in the contract, possibly because of his claim that its use will reduce the number or the length of the piles needed. If both the contract system and the substituted system provide for payment on a basis of measured work, then whether or not a saving is ultimately achieved will never be known. Thus it is necessary to distinguish between a variation requested by the contractor and allowed by the engineer which does not affect the evaluation of the work (for example, a simple substitution of a different material from that specified) and one in which the engineer adopts, at the contractor's suggestion, an alternative form of construction. In the latter case, it is usually necessary to prepare a supplementary bill of quantities to cover the varied work, and to define those items in the original bill which are superseded.

Clause 51(2) includes sensible provisions (set out in detail in clause 2(6)) intended to ensure that a written record is available of all variations ordered. Where either the engineer or the contractor complies with the sub-clause, and provided that the written order or the contractor's confirmation of the engineer's oral order is clear and unambiguous, the likelihood of disputes arising as to whether or not orders were given or as to their content is greatly reduced. A claim by the engineer that the contractor's written confirmation, although not contradicted by the engineer at the time, does not truly represent what he ordered orally must automatically fail, in that the written confirmation is deemed to be an order in writing by the engineer.

Clause 51(3) defines the status of variation orders: namely that they do not vitiate or invalidate the contract although that their value affects the contract price (except to the extent that such a variation results from the contractor's default). The exception was not expressed in the Fifth Edition.

Clause 51(4) provides that no written order is needed where the variation results simply from the quantity of work required differing from that given in the bill of quantities. Such a variation may, however, warrant an adjustment, either upwards or downwards, in the rate or price for the item: see clause 56(2). Similarly, a variation order which alters the quantity, but not the type or quality, of work required may warrant an adjustment in the price or rate: see clause 52(2).

4.2 *Evaluation of ordered variations*

Clause 52 of the ICE Conditions states:

(1) The value of all variations ordered by the Engineer in accordance with Clause 51 shall be ascertained by the Engineer after consultation with the Contractor in accordance with the following principles.

(a) Where work is of similar character and executed under similar conditions to work priced in the Bill of Quantities it shall be valued at such rates and prices contained therein as may be applicable.

(b) Where work is not of a similar character or is not executed under similar conditions or is ordered during the Defects Correction Period the rates and prices in the Bill of Quantities shall be used as the basis for valuation so far as may be reasonable failing which a fair valuation shall be made.

Failing agreement between the Engineer and the Contractor as to any rate or price to be applied in the valuation of any variation the Engineer shall determine the rate or price in accordance with the foregoing principles and he shall notify the Contractor accordingly.

(2) If the nature or amount of any variation relative to the nature or amount of the whole of the contract work or to any part thereof shall be such that in the opinion of the Engineer or the Contractor any rate or price contained in the Contract for any item of work is by reason of such variation rendered unreasonable or inapplicable either the Engineer shall give to the Contractor or the Contractor shall give to the Engineer notice before the varied work is commenced or as soon thereafter as is reasonable in all the circumstances that such rate or price should be

varied and the Engineer shall fix such rate or price as in the circumstances he shall think reasonable and proper.

(3)   The Engineer may if in his opinion it is necessary or desirable order in writing that any additional or substituted work shall be executed on a daywork basis in accordance with the provisions of Clause 56(4).

The phrase 'after consultation with the Contractor' emphasises the important principle that wherever possible rates and/or prices for varied work should be agreed between the engineer and the contractor. Where the contractor does not submit a proposed rate for the varied work the engineer should invite him to do so and should, if necessary, discuss that proposal with a view to agreeing any adjustment the engineer may believe to be necessary. At such discussions, the contractor should provide the basis of his evaluation and should be willing to justify his proposals.

Sub-clause (1) defines three bases of evaluation:

- Work that is both similar in character and executed under similar conditions to work included in the bill of quantities is to be valued at such rates and prices in the contract 'as may be applicable'.
- For work that is not similar in character or is not executed under similar conditions or is ordered during the defects correction period the rates in the bill of quantities must be used as a basis of valuation 'so far as may be reasonable'.
- Where the use of contract rates fails a 'fair valuation' must be made.

Impliedly, the first two bases include elements for profit, since the rates and prices in the bill of quantities must in general include such elements. Any cost of prolongation or disruption associated with variations must be recovered under clause 52, unless the contractor can rely on other clauses.

4.2.1   Application of contract rates

The simple application of rates in the bill of quantities to work of a similar character executed under similar conditions should not in itself cause any difficulty. Contention may however sometimes arise as to whether the conditions are similar, for example the method of handling concrete may have to be changed to some more costly method where the varied work is not within the range of the contractor's tower crane. In such a case the contractor should of course submit his claim for the additional costs resulting from the less accessible location of the varied work, with evidence in support of the claim.

In some cases, where a change in the quantity of work executed under an item is such as to 'render [the rate] unreasonable or inapplicable' – irrespective of the amount of the variation or its proportion to the billed quantity – sub-clause (2) will apply. That sub-clause does not however provide a means whereby pricing errors or intentional distortion of rates included in the contract may be corrected. This is because such rates are or may be already unreasonable, even before the variation is ordered – it is not a case of their being rendered unreasonable by the variation.

While one can perhaps sympathise with a contractor who, having erroneously under-priced an item in the bill, finds the quantity against that item to be vastly increased, sub-clause (2) provides him with a remedy only if and to the extent that he can show the rate to have been rendered unreasonable by a variation and not by his own error. From the employer's viewpoint the error in pricing might have been a deliberate distortion intended to take advantage of a suspected error in the quantity. (See section 4.4 below for errors in pricing.)

Where an inadequate rate for extra work, or an excessive deduction in respect of omitted work, results from such deliberate distortions the contractor has only himself to blame. He has elected to gamble on the final quantities of work, and has lost.

The way of avoiding the opposite situation, namely where the contractor's gamble on final quantities succeeds at the employer's expense, lies in the hands of the engineer. He should be better able than the contractor to foresee the final quantities; if he does so and enters them correctly in the bill of quantities there is no opportunity for the contractor to profit from the engineer's errors. Furthermore, where the engineer notices rates in a tender that appear to have been distorted he should, before recommending acceptance of that tender, consider carefully the effect those distortions might have on any likely variations of quantity. Such consideration could possibly lead him to the conclusion that acceptance of the lowest tender will not result in the lowest final cost to the employer, and hence that some other tender should be accepted.

It does, of course, occasionally happen that the engineer deliberately increases quantities of types of work of which the final quantity is uncertain in order to provide a hidden contingency sum. A favourite example of this is an 'extra over' item for rock excavation. A contractor who is aware of this approach may be tempted to reduce his rate for the item in question, perhaps even to zero, and to make up the amount so deducted from his estimate by increasing rates or prices for other items in his tender. Here again, the remedy lies in the hands of the engineer, who should not distort measured or properly estimated quantities in order to provide for contingencies, but should allow for contingencies in items so described.

4.2.2   Work of a different character or under different conditions

In applying rates in the bill of quantities to work of a different character or work executed under different conditions the starting point must always be an examination of the build-up of the rate for work corresponding most closely to the varied work. Contractors are often secretive about their techniques in estimating. They have the right to expect that the engineer will not disclose to unauthorised persons information provided for purposes of agreeing rates, but they must accept the engineer's entitlement to information needed to support their claim. Refusal to provide such information can only result in the engineer fixing a rate at the minimum he can see to be justified without, for example, adding any fixed percentage that the contractor may have added to his contract rates as a contribution towards overheads and profit.

4.2.3   A fair valuation

Where the bill of quantities contains no rate from which a rate for the varied work may be derived, a 'fair valuation' must be made. The extent to which a 'fair valuation' should have regard to the rates and prices that have not been applied is unclear. It is probable that the 'fair valuation' must have regard to the way in which rates in the contract have been estimated, even though there may be no rate for work that even remotely resembles the varied work. This is because a 'fair valuation' must represent, as closely as possible, the rate the contractor would have inserted had the work in question been included in the bill of quantities. Hence it is reasonable to expect, for example, that the contractor's build-up of his labour costs for various classes of labour will be applied to the varied work, as will his rates for plant. In the case of materials, the cost mark-up used generally in the contractor's estimate should be applicable to the materials required for the varied work, as should relevant allowances for waste.

In each case it is open to the contractor to contend that such basic principles require adjustment – for example, that the labour and plant rates for work that is especially dependent upon the favourable weather conditions during which the billed work was planned to be carried out should be enhanced to allow for the likely increase in the proportion of 'rained-off' time if the varied work has to be carried out during the winter months.

4.2.4 Rates fixed by engineer where affected by quantity variations

Clause 52(2) empowers the engineer to fix varied rates to allow for variations in quantity. It is a popular belief that within some predetermined percentage limit – often quoted as 10% – variations of quantity do not justify variation of the relevant rate, while outside that limit a varied rate becomes an entitlement. However that is not what the contract provides. Entitlement to a varied rate depends upon the party seeking the variation – which may be either the contractor or, through his engineer, the employer – being able to show that the contract rate is *by reason of a variation in the quantity* rendered unreasonable or inapplicable.

Hence it is clear that the provision does not allow for adjustment of a rate that is unreasonable before the variation is ordered, such as a rate that has been distorted by the contractor in an attempt to gain from expected variations in quantity, or a rate that is simply erroneous. The provision will, however, apply where the cost of a lump-sum element of the work has to be spread over a number of items, and where that number is varied.

For example, where a contractor's rate for shuttering includes an element for purpose-made shutters which are estimated at the time of tendering to be used x times, a reduction in the number of uses resulting from a variation must justify a notice by the contractor under sub-clause 52(2) that the rate is rendered unreasonable or inapplicable by reason of the variation. Similarly an increase in the number of uses would justify a notice by the engineer under the sub-clause that the rate should be reduced. Again, where the contract requires the provision of a quantity of filling material, the exact quantity of which is available from a nearby borrow pit, an increase in the quantity necessitating the use of a more remote pit for the additional quantity would justify an enhancement of the rate in respect of the additional quantity to allow for the increased haulage costs.

4.2.5 Daywork

Clause 52(3) provides that the engineer may order the execution of varied work as daywork, and refers to clause 56(4), which provides as follows:

(4)   Where any work is carried out on a daywork basis the Contractor shall be paid for such work under the conditions and at the rates and prices set out in the daywork schedule included in the Contract or failing the inclusion of a daywork schedule he shall be paid at the rates and prices and under the conditions contained in the 'Schedules of Day-works carried out incidental to Contract Work' issued by the Civil Engineering Contractor's Association (formerly issued by the Federation of Civil Engineering Contractors) current at the date of execution of the daywork.

The Contractor shall furnish to the Engineer such records receipts and other documentation as may be necessary to prove amounts paid and/or costs incurred. Such returns shall be in the form and delivered at the times the Engineer shall direct and shall be agreed within a reasonable time.

Before ordering materials the Contractor shall if so required submit to the Engineer quotations for the same for his approval.

Although some major contracts include schedules of dayworks it is more usual for the engineer to rely upon the *Schedules of Dayworks Carried out incidental to Contract Work,* now published by the Civil Engineering Contractors Association ('CECA').

Where records of daywork are properly maintained, submitted daily to the engineer, and checked and signed by him, the system has the merits of providing an agreed record from which the final cost may be determined without difficulty (albeit with some considerable expenditure of clerical labour) and of obviating contention as to rates. However, daywork has the major disadvantage of removing the incentive, both to the contractor and to his employees, to achieve a good rate of productivity. Experienced workmen on a site are soon aware when work is to be measured as daywork, and in many cases see such work as a welcome opportunity to relax between periods of intense activity on piecework or on work for which productivity bonuses provide the incentive.

The CECA dayworks schedules include a definition of the incidental expenses covered by the percentage addition (currently 148%) to 'amount of wages', which term is also defined. Contention may, however, arise where men are employed for part of a day on daywork, in which case items such as daily travelling allowances must be apportioned between time spent on daywork and on other duties. The engineer should also be careful to ensure that items such as site supervision, which are expressly included in the percentage addition to the amount of wages, are not duplicated by the inclusion of hours worked by supervisory staff, such as non-working foremen, on the daywork sheets.

4.3 Notice of claims

Clause 52(4) of the ICE Conditions states:

> (4) (a) If the Contractor intends to claim a higher rate or price than one notified to him by the Engineer pursuant to sub-clauses (1) and (2) of this Clause or Clause 56(2) the Contractor shall within 28 days after such notification give notice in writing of his intention to the Engineer.
> (b) If the Contractor intends to claim any additional payment pursuant to any Clause of these Conditions other than sub-clauses (1) and (2) of this Clause or Clause 56(2) he shall give notice in writing of his intention to the Engineer as soon as may be reasonable and in any event within 28 days after the happening of the events giving rise to the claim. Upon the happening of such events the Contractor shall keep such contemporary records as may reasonably be necessary to support any claim he may subsequently wish to make.
> (c) Without necessarily admitting the Employer's liability the Engineer may upon receipt of a notice under this Clause instruct the Contractor to keep such contemporary records or further contemporary records as the case may be as are reasonable and may be material to the claim of which notice has been given and the Contractor shall keep such records. The Contractor shall permit the Engineer to inspect all records kept pursuant to this Clause and shall supply him with copies thereof as and when the Engineer shall so instruct.
> (d) After the giving of a notice to the Engineer under this Clause the Contractor shall as soon as is reasonable in all the circumstances

send to the Engineer a first interim account giving full and detailed particulars of the amount claimed to that date and of the grounds upon which the claim is based. Thereafter at such intervals as the Engineer may reasonably require the Contractor shall send to the Engineer further up to date accounts giving the accumulated total of the claim and any further grounds upon which it is based.

(e) If the Contractor fails to comply with any of the provisions of this Clause in respect of any claim which he shall seek to make then the Contractor shall be entitled to payment in respect thereof only to the extent that the Engineer has not been prevented from or substantially prejudiced by such failure in investigating the said claim.

(f) The Contractor shall be entitled to have included in any interim payment certified by the Engineer pursuant to Clause 60 such amount in respect of any claim as the Engineer may consider due to the Contractor provided that the Contractor shall have supplied sufficient particulars to enable the Engineer to determine the amount due. If such particulars are insufficient to substantiate the whole of the claim the Contractor shall be entitled to payment in respect of such part of the claim as the particulars may substantiate to the satisfaction of the Engineer.

Clause 52(4)(a) envisages a situation in which the engineer has fixed a rate under sub-clause (1) or (2), presumably in any of the three ways defined in sub-clause (1), which the contractor considers to be insufficient. This provision has the merit of drawing attention to any potential dispute at an early date and while it may still be possible for the engineer to cancel or to modify his variation order from which the rate originated.

Sub-clauses (4)(b) and (c) are important in that they relate to claims arising under *any* clause of the contract. For this reason, their inclusion in a clause relating solely to variations and not to other matters that may give rise to claims is misleading. The provision for maintaining 'contemporary' records is in the interests of both parties in that disputes on matters of *quantum* are or should be avoided where proper records are maintained.

Whether or not the engineer so orders under clause 52(4)(c), a prudent contractor will keep records of his costs wherever there is a possibility of a claim and will seek the RE's agreement to those records at regular intervals.

The practice of some engineers and REs in refusing to sign records of labour, plant and materials where the existence of a claim is disputed is unhelpful to the resolution of any future dispute and is unlikely to further the employer's interests. If the only records available in an arbitration are those maintained by the contractor, then the employer may be in great difficulty in attempting to refute those records. This is especially the case where the contractor can demonstrate that he sought to obtain, but was refused, the engineer's agreement at the time of the events in question. Even the maintenance of a separate record by the engineer is not decisive. An arbitrator confronted with conflicting evidence of factual matters may have difficulty in deciding between them. If the contractor can show that he sought to agree contemporaneous records but was prevented from doing so by the engineer that fact is likely to weigh heavily in the contractor's favour.

Sub-clause (4)(d) reaffirms the standard requirement that records must be provided as soon as it is practicable to do so. In many cases the full extent of a claim will not become known for some time after the events from which it arises have become known, and in such cases the contractor can only provide, with minimum delay, the information as it becomes available.

Sub-clause (4)(e), while softening the somewhat draconian provisions of the Fourth Edition of the Conditions, should not be taken to indicate that notices are a formality that may be ignored without incurring a penalty. Notice of a claim alerts the engineer to its existence and gives him an opportunity not only to record details that may be needed for evaluation purposes but also to seek some means of avoiding or limiting the circumstances which give rise to the claim.

Thus, for example, it may be possible to re-route an underground pipe or cable so as to avoid a local obstruction, or to obtain additional filling required because of a variation from a source known to the engineer but not to the contractor, or to vary the design of foundations where unforeseen subsoil conditions make such variation economic.

Sub-clause (4)(f) clarifies the stage at which payment in respect of claims becomes due. It emphasises the advantages to be gained by the contractor in providing full details of evaluation at the earliest possible date. Even though the engineer may reject the claim, an arbitrator may subsequently allow it; and his award of interest must in general be related to his finding of when the amount he awards ought to have been paid. That date will depend upon when the contractor provided full details of the facts from which the claim arises and substantiation of its evaluation. The provisions of clause 52(4) are discussed in more detail in Chapter 7.

4.4 Errors in bills of quantities

Clause 55 of the ICE Conditions states:

(1) The quantities set out in the Bill of Quantities are the estimated quantities of the work but they are not to be taken as the actual and correct quantities of the Works to be executed by the Contractor in fulfilment of his obligations under the Contract.

(2) Any error in description in the Bill of Quantities or omission therefrom shall not vitiate the Contract nor release the Contractor from the execution of the whole or any part of the Works according to the Drawings and Specification or from any of his obligations or liabilities under the Contract. Any such error or omission shall be corrected by the Engineer and the value of the work actually carried out shall be ascertained in accordance with Clause 52. Provided that there shall be no rectification of any errors omissions or wrong estimates in the descriptions rates and prices inserted by the Contractor in the Bill of Quantities.

Clause 55(1) clarifies the status of the bill of quantities and by express denial refutes any claim that might otherwise be made of the existence of an implied warranty that the billed quantities are accurate. Clause 55(2) reaffirms the principle of clause 51, namely that the engineer has the authority, and in some cases the duty, to impose variations unilaterally, but with the proviso that the value of variations is taken into account in determining the value of the contract.

Errors of description cover, for example, incorrect descriptions of the quality of such items as concrete, timber and steel, or of the finish required to surfaces of those materials. These errors are dealt with in accordance with the principles defined in clause 52.

Errors in billed quantities are covered by sub-clause (2). Clearly, rectification of a gross error – such as the inclusion of too many or too few noughts in a quantity – would sometimes result in a variation of the rate under clause 56(2) or, where a variation is ordered, clause 52(2). In some cases the existence of such a gross error in billed quantities is brought to the engineer's attention during the tendering period. In such a case, the engineer should ensure that all tenderers are notified of the correction.

Regrettably this is not always done and in consequence tenderers are sometimes wary of pointing out errors. The tenderer who does so puts himself at a disadvantage in comparison with his competitors who, if the engineer does not issue a correction generally and if they are awarded the contract, may be able to claim additional payment when the error comes to be corrected. The tenderer who pointed out the error can make no such claim – he is known to have been aware of the error at the time he prepared his tender and impliedly to have allowed for it in his rates.

The final provision of sub-clause (2) frequently gives rise to misunderstanding in cases where an item that has been incorrectly priced by the contractor is subjected to a variation in quantity. Clearly that incorrect price, the extension of which formed an element of the tender total, must be applied to the quantity as billed, so that without the variation there would be no problem. But if the rate is grossly excessive and the quantity is increased, is the contractor entitled to a gross increase in his profit? Or, if the quantity is reduced, must he accept a gross reduction in his profit?

The answer in both cases seems to be in the affirmative. There is no means of distinguishing between a rate that has been distorted deliberately to take advantage of an expected variation of quantity or for any other reason, and one that has been the subject of an error. In both cases the parties to the contract have agreed to the rate, however unreasonable it may be, and they must – unless they negotiate some variation to the contract – abide by that rate. Neither clause 52(2) nor clause 56(2) is applicable, because it is not the variation that renders the rate unreasonable – the rate is already unreasonable because of the error or deliberate distortion.

However, an exception arises where both parties err in making their agreement. For example, where an item for concrete 100 mm thick is inadvertently billed as being measured by the square metre instead of by the cubic metre, and neither party notices the error, the contractor would find his rate to be ten times the rate he intended, and that it is applied to a quantity that is also multiplied by ten. That situation would, it is submitted, be covered by the general rule of law that a contract made as a result of an operative mistake is void.

4.5   Method of measurement

Clause 57 of the ICE Conditions states:

> Unless otherwise provided in the Contract or unless general or detailed description of the work in the Bill of Quantities or any other statement clearly shows to the contrary the Bill of Quantities shall be deemed to have been prepared and measurements shall be made according to the procedure set out in the 'Civil Engineering Standard Method of

Measurement Second Edition 1985' approved by the Institution of Civil Engineers and the Federation of Civil Engineering Contractors in association with the Association of Consulting Engineers or such later or amended edition thereof as may be stated in the Appendix to the Form of Tender to have been adopted in its preparation.

Clause 57 of the Sixth Edition makes it clear that the 1985 revision of the *Standard Method of Measurement* document – which does not depart radically from its predecessor – now applies unless otherwise provided.

In the case of road and bridge works for the Department of Transport it is usual for clause 57 and the Appendix to the form of tender to be amended so as to refer to the Department's *Method of Measurement for Road and Bridge Works* (MMRB), the second edition of which was published in 1977.

The provision, through the CESMM or MMRB, of definitions of the work included in the description of each item of the bill of quantities reduces the risk that disputes may arise as to the meaning of such descriptions. Although the clause performs the valuable service of standardising such definitions, a provision has been introduced in the Sixth Edition that detracts from that service – namely the inclusion of the words 'or any other statement' in line 2 of the clause.

Under the Fifth Edition any deviation from the Standard Method of Measurement had to be expressly shown in the bill of quantities. Thus an estimator, when pricing individual items, was relieved of the task of searching through the specification, drawings and other contract documents for any such deviation, which could of course affect his pricing. Regrettably that sensible provision no longer applies. The new wording of the Sixth Edition allows exceptions to be made in 'any other statement', which phrase would presumably allow them to be included in the specification or even perhaps on the drawings.

It may perhaps be arguable that such a statement which, although clear in itself, is not to be found where an estimator might expect to find it – namely in the bill of quantities – is not one which 'clearly shows to the contrary'. If such a construction is tenable then the damage resulting from the change of wording may be minimal; but it is in our view unfortunate that an element of uncertainty should have been introduced into a clause that was, in the Fifth Edition, satisfactory.

Although clause 57 is sometimes criticised for providing a source of claims that definitions in the CESMM or MMRB do not cover all of the work required to be performed, such claims can arise only where the bill of quantities fails to comply with the standard. Without such a clause it is likely that many more claims would arise from the lack of a precise definition of item coverage.

4.6   *Variations under the Minor Works Form*

Clause 2.3(a) of the Minor Works Form provides:

> The Engineer shall have power to give instructions for:
> (a)   any variation to the Works including any addition thereto or omission therefrom ...

Variations fall to be valued under clause 6, which provides as follows:

> 6.1   If the Contractor carries out additional works or incurs additional cost including any cost arising from delay or disruption to the progress of the Works as a result of any of the matters referred to in paragraphs (a) (b) (d) (e) or (f) of Clause 4.4 the Engineer shall certify and the Employer shall pay to the Contractor such additional sum as the Engineer after consultation with the Contractor considers fair and reasonable. Likewise the Engineer shall determine a fair and reasonable deduction to be made in respect of any omission of work.
> 6.2   In determining a fair and reasonable sum under Clause 6.1 for additional work the Engineer shall have regard to the prices contained in the Contract.

A variation order may also give rise to an extension of time for completion under clause 4.4.

Where the variation orders additional work, the engineer is to 'have regard to the prices contained in the Contract', i.e. the priced bill, and the Schedule of Rates or Daywork Schedules, as appropriate, when valuing the work. The reference to consultation with the contractor is important.

CHAPTER FIVE
DELAYS, LIQUIDATED DAMAGES AND EXTENSIONS OF TIME

5.1 Introduction

Failure to complete the works on time is a breach of contract on the part of the contractor. The normal remedy for breach is a claim for damages, which requires careful calculation. Damages are awarded so as to place the successful claimant in the same position as if the contract had been performed. The claimant is required to prove his loss. This can be an onerous task. How can the damage caused by the late completion of a road or bridge be proved? It is also very uncertain. The potential losses may be so huge that no contractor would be prepared to enter into a contract, knowing that late completion may result in almost limitless claims for damages.

In order to deal with these difficulties, most forms of construction contract including both the ICE Conditions and the Minor Works Conditions provide for the payment of liquidated damages if the contractor completes late. A liquidated damages clause is an agreement for the payment of agreed damages for potential breaches of contract – in this case, the breach of late completion. The sum of money agreed as liquidated damages by the parties is recoverable whether or not, in the event, the employer can prove that he has in fact suffered loss as a result of the breach: *BFI Ltd* v. *DCB Integration Systems Ltd* (1987). Indeed, the very point of specifying liquidated damages is that they are recoverable without the need to prove loss.

The clause also acts as a limitation on the contractor's liability for the breach of late completion. It can be seen as exhaustive of the employer's rights in this respect. In *Temloc Ltd* v. *Errill Properties Ltd* (1987), a building contract case, the contract appendix entry of the amount of liquidated damages was entered as '£NIL'. Third parties made claims against the employer arising out of delay in completion. The Court of Appeal held that the employer was not entitled to claim *any* relief for delay in completion by the contractor on the basis that the provision for liquidated damages was an exhaustive remedy.

Lord Justice Nourse said:

> 'I think it clear, both as a matter of construction and as one of common sense, that if ... the parties complete the relevant parts of the appendix ... then that constitutes an exhaustive agreement as to the damages which are ... payable by the contractor in the event of his failure to complete the works on time.'

Whether or not a liquidated damages clause is an exhaustive remedy is, however, always a matter of construction. If it is held to be exhaustive, the employer is not entitled to elect to claim general damages instead, but *Temloc Ltd* v. *Errill Properties Ltd* was distinguished by the Supreme Court of New South Wales in *Baese Pty Ltd* v. *R A Bracken Building Pty Ltd* (1989) and if, under a JCT contract (or the Minor Works Conditions), the appendix entry were to be left blank it is thought that general damages would be recoverable. The position is different under the Sixth Edition of the ICE Conditions since, as will be seen, clause 47(4)(b) expressly provides that where 'nil' is inserted in the Appendix 'or a figure is omitted therefrom' then no damages shall be payable.

Liquidated damages provisions are normally linked with an extension of time clause empowering the engineer to extend the period for completion. Provisions for extension of time are necessary because without them the contractor would be under a strict duty to complete on time unless he was hindered or prevented from doing so by acts or breaches of the employer or of those for whom the employer is responsible in law: *Percy Bilton Ltd.* v. *Greater London Council* (1982).

The consequence of this is that a liquidated damages clause will fail if extensions are not grantable under the contract and circumstances arise which amount to an act of interference or prevention by the employer. This includes variations, late information and so on. The liquidated damages clause would then cease to apply even to subsequent delays that were the contractor's responsibility and, once lost, the right to liquidated damages cannot be revived. It is for this reason that extension of time clauses are often said to be for the employer's benefit.

This follows from *Peak Construction (Liverpool) Ltd* v. *McKinney Foundations Ltd* (1970) where there was a delay of 58 weeks for remedial work that took only six weeks. The greater part of the delay was caused by the employer's failure to make a decision on what was to be done. The Court of Appeal held that the employer could not recover liquidated damages. He loses his right to recover liquidated damages if any part of the delay is caused by any act or conduct of his for which no specific extension is provided for by, and granted under, the contract.

The Court of Appeal said that if the liquidated damages clause fails, the employer 'is left to prove such general damages as he may have suffered', that is, unliquidated damages determined by the court after the breach. It is not clear, however, whether the claim for unliquidated damages has a ceiling on it equal to the amount of the defunct liquidated damages. The point was left undecided by the Court of Appeal in *The Rapid Building Group Ltd* v. *Ealing Family Housing Association Ltd* (1984). If general damages in excess of the defunct liquidated damages are claimable, that would be a bonus for the employer. The better view is probably that there is a ceiling. *Keating on Building Contracts*, 6th edn, 1995, p252 puts the matter thus:

> 'It is thought that where the cause of the unenforceability of the liquidated damages provision is the employer's act, the Court would not permit the employer to recover greater damages than the liquidated damages which would, if enforceable, have been payable.'

5.2 Liquidated damages or penalty?

Throughout the civil engineering industry, it is common to refer to a contract provision for liquidated damages as 'the penalty clause'. This is misleading. Liquidated damages are recoverable at law whereas a 'penalty' is not. A penalty is an extravagant sum of money in comparison to the greatest possible loss which could be suffered by the employer. If the amount specified is exorbitant, it will constitute a penalty and be irrecoverable.

To avoid being a penalty the rate stated in the contract must be a genuine pre-estimate of the employer's likely loss or a lesser sum. How the parties describe the sum is irrelevant, and so the statement in clause 47(3) of the Sixth Edition that 'all sums payable by the contractor to the employer ... shall be paid as liquidated damages ... and not as a penalty' is of no effect unless the amount *is* a genuine pre-estimate of the likely damage caused by any delay or a lesser amount, i.e. liquidated damages.

The engineer must, therefore, assess the likely loss to the employer if the works are not completed on time, and then reduce this figure to a daily or weekly rate. It does not matter if the estimate turns out, in fact, to be a poor one or if a precise pre-estimate is not possible.

In practice, the sums agreed in the civil engineering industry are comparatively small in relation to the employer's potential loss because a realistic figure may well act as a disincentive to potential tenderers or result in over-pricing. Normally, the figure assessed will be reduced to a figure that is commercially attractive and often an overall percentage of the estimated final contract value is taken as a ceiling – but a figure should not be plucked out of the air.

In some cases estimating the likely loss to the employer is inherently difficult, especially in public works contracts. Such a difficulty is often a very good reason for adopting liquidated damages because they are recoverable without the employer being put to proof of loss, which could be extremely difficult, complex and expensive.

It is immaterial whether the employer's actual loss is greater or less than the agreed figure. He can recover only the stipulated sum, and he is entitled to liquidated damages at the rate agreed even if the breach causes no loss at all to him. The genuineness of the pre-estimate is to be judged at the time the contract is made and not at the time of the breach. This is a point that contractors often find hard to grasp.

The essential point is that a genuine pre-estimate should be made from data that is capable of verification if challenged. In the public sector, some kind of formula approach is often used. In the related field of building contracts, the Society of Chief Quantity Surveyors in Local Government has published a comprehensive report on the calculation of liquidated damages, which suggested a basis on which the calculation might be made:

- At the anticipated completion date it is assumed that 80% of the total capital cost of the scheme will have been advanced and that interest at the current rate is being paid. If an interest rate of 12% is assumed, the cost of interest will be:

$$\frac{80\% \times 12\%}{52} = 0.185\% \text{ of the contract price per week}$$

- Professional fees likely to be incurred as a result of delay are added.
- Other specific costs likely to be incurred, such as temporary accomodation charges, are added if appropriate, together with an allowance for increased costs payable by way of fluctuations.

This approach is permissible provided a genuine calculation is made and a reasonable overall figure is assessed from verifiable data. Indeed there appear to be no reported cases in which a sum stated in a civil engineering contract has been held to be a penalty because of its amount. Provided a genuine pre-estimate is made by the engineer or his staff, and the contractual machinery for extensions of time is properly operated, it is improbable that a contractor could successfully challenge the agreed figure under a contract made on the ICE or Minor Works conditions. The contractor's agreement to the figure is signified by his entering into the contract.

This commercial approach was approved by the Privy Council in 1993 in what has become the leading case on the enforcement of liquidated damages in recent years, *Philips Hong Kong Ltd* v. *The Attorney General of Hong Kong* (1993). Philips had contracted to carry out the design and installation of a computerised supervisory system for a road network in Hong Kong. There were several contracts let to different contractors for various parts of the network, with key dates for various parts of each contract which had to be met in order to provide an overall co-ordinated programme for the works as a whole. If the contractor failed to meet any of these dates, liquidated damages were payable in respect of each key date at various daily rates, and further liquidated damages were payable if the overall completion date was missed. The maximum figure that Philips might have to pay in liquidated damages was potentially very high. Philips sought a declaration of the court that the liquidated damages provision was unenforceable because it was in reality a penalty.

Philips were successful at first instance in Hong Kong, but the Court of Appeal in Hong Kong reversed the High Court decision. The Privy Council, as the ultimate Court of Appeal, dismissed Philips' appeal against that decision, on the basis that it was reluctant to interfere in a commercial contract, where the liquidated damages figure had been calculated with some care and agreed. It effectively reserved its position in cases where there is serious inequality of bargaining power, and an unreasonable figure has been imposed in the contract, but made it clear that such cases would be rare.

In the unlikely event of a contractor satisfying an arbitrator or the court that the agreed figure is an unenforceable penalty, this would not mean that he was released from liability for the financial consequences of his breach of contract. The employer would be left with the right to claim unliquidated damages under the general law, assessed by the arbitrator or the court and subject to proof of loss: *Peak Construction (Liverpool) Ltd* v. *McKinney Foundations Ltd* (1970). Possibly, the general damages then recoverable would be subject to a ceiling on the amount, as already indicated.

A liquidated damages clause is more likely to fail because the contract provisions for extension of time have not been properly exercised by the engineer, or where there has been some delay that is the responsibility of the employer and which is not covered by the extension of time clause as was the situation in the old case of *Wells* v. *Army & Navy Co-operative Society Ltd* (1902). There the relevant clause allowed for extensions for various matters causing delay including 'other causes beyond the contractor's control'. A three-month extension was allowed for delays caused by sub-contractors, the contract having overrun by a year. Other breaches that were the employer's responsibility were established, including failure to give possession of the site and late information, drawings and so on. The High Court held that the words 'other causes beyond the contractor's control' did not cover delays caused by the employer's defaults. Liquidated damages were not therefore recoverable.

Where the liquidated damages clause fails in such a case there is no fixed date for completion. The contractor's obligation then is to complete the works 'within a reasonable time' – in legal parlance time is 'at large'. The question as to what is a 'reasonable time' is a question of fact and in deciding what is a reasonable time all the circumstances must be taken into consideration. The contractor's failure to complete within a reasonable time would be a breach of contract for which the employer would have a remedy in unliquidated damages.

5.3   Liquidated damages – the contract provisions

5.3.1   ICE Conditions

Clause 47 of the ICE Conditions provides:

> (1) (a) Where the whole of the Works is not divided into Sections the Appendix to the Form of Tender shall include a sum which

represents the Employer's genuine pre-estimate (expressed per week or per day as the case may be) of the damages likely to be suffered by him if the whole of the Works is not substantially completed within the time prescribed by Clause 43 or by any extension thereof granted under Clause 44 or by any revision thereof agreed under Clause 46(3) as the case may be.

(b)   If the Contractor fails to complete the whole of the Works within the time so prescribed he shall pay to the Employer the said sum for every week or day (as the case may be) which shall elapse between the date on which the prescribed time expired and the date the whole of the Works is substantially completed.

Provided that if any part of the Works is certified as complete pursuant to Clause 48 before the completion of the whole of the Works the said sum shall be reduced by the proportion which the value of the part so completed bears to the value of the whole of the Works.

(2)   (a)   Where the Works is divided into Sections (together comprising the whole of the Works) which are required to be completed within particular times as stated in the Appendix to the Form of Tender sub-clause (1) of this Clause shall not apply and the said Appendix shall include a sum in respect of each Section which represents the Employer's genuine pre-estimate (expressed per week or per day as the case may be) of the damages likely to be suffered by him if that Section is not substantially completed within the time prescribed by Clause 43 or by any extension thereof granted under Clause 44 or by any revision thereof agreed under Clause 46(3) as the case may be.

(b)   If the Contractor fails to complete any Section within the time so prescribed he shall pay to the Employer the appropriate stated sum for every week or day (as the case may be) which shall elapse between the date on which the prescribed time expired and the date of substantial completion of that Section.

Provided that if any part of that Section is certified as complete pursuant to Clause 48 before the completion of the whole thereof the appropriate stated sum shall be reduced by the proportion which the value of the part so completed bears to the value of the whole of that Section.

(c)   Liquidated damages in respect of two or more Sections may where circumstances so dictate run concurrently.

(3)   All sums payable by the Contractor to the Employer pursuant to this Clause shall be paid as liquidated damages for delay and not as a penalty.

(4)   (a)   The total amount of liquidated damages in respect of the whole of the Works or any Section thereof shall be limited to the appropriate sum stated in the Appendix to the Form of Tender. If no such limit is stated therein then liquidated damages without limit shall apply.

(b)   Should there be omitted from the Appendix to the Form of Tender any sum required to be inserted therein either by sub-clause (1)(a) or by sub-clause (2)(a) of this Clause as the case may be or if any such sum is stated to be 'nil' then to that extent damages shall not be payable.

(5)   The Employer may

(a)   deduct and retain the amount of any liquidated damages becoming due under the provision of this Clause from any sums due or which become due to the Contractor or

(b)   require the Contractor to pay such amount to the Employer forthwith.

If upon a subsequent or final review of the circumstances causing delay the Engineer grants a relevant extension or further extension of time the Employer shall no longer be entitled to liquidated damages in respect of the period of such extension.

Any sum in respect of such period which may already have been recovered under this Clause shall be reimbursed forthwith to the Contractor together with interest compounded monthly at the rate provided for in Clause 60(7) from the date on which such sums were recovered from the Contractor.

(6)   If after liquidated damages have become payable in respect of any part of the Works the Engineer issues a variation order under Clause 51 or adverse physical conditions or artificial obstructions within the meaning of Clause 12 are encountered or any other situation outside the Contractor's control arises any of which in the Engineer's opinion results in further delay to that part of the Works

(a)   the Engineer shall so inform the Contractor and the Employer in writing and

(b)   the Employer's entitlement to liquidated damages in respect of that part of the Works shall be suspended until the Engineer notifies the Contractor and the Employer in writing that the further delay has come to an end.

Such suspension shall not invalidate any entitlement to liquidated damages which accrued before the period of delay started to run and any monies deducted or paid in accordance with sub-clause (5) of this Clause may be retained by the Employer without incurring liability for interest thereon under Clause 60(7).

Clause 47 provides for the payment of liquidated damages by the contractor if he fails to complete the works on time. There are significant differences between clauses 47 in the Fifth and Sixth Editions. The provision will also operate as a limit on the contractor's liability for damages for late completion where, as is usually the case, a lesser sum than a genuine pre-estimate of the employer's likely loss is entered in column 1 of the Appendix entry.

Delay of the whole of the works

Clause 47(1)(a) deals with liquidated damages for the *whole* of the works and is straightforward in its operation. The major point to note is that the Appendix entry must be completed correctly and consistently with the printed provision. If the Appendix entry is left blank or the sum is stated to be 'nil' then (clause 47(4)(b)) 'to that extent damages shall not be payable'.

Liquidated damages are payable at the prescribed weekly or daily rate should the contractor fail to complete the whole of the works within the prescribed contract period or any extension of time granted by the engineer under clause 44. The period of payment is 'for every week or day (as the case may be) which shall elapse between the date on which the prescribed time expired and the date the whole of the Works is substantially completed', i.e. substantial completion, as envisaged by clauses 43 and 48.

Clause 43 sets out the contractor's obligation to ensure that the works are 'substantially completed' within 'the time ... stated (or such extended time as may be allowed under Clause 44)' and clause 48 enables the Certificate of Substantial Completion to be issued provided the work has been substantially completed and the contractor gives an undertaking to finish any outstanding work.

The proviso to clause 47(1)(b) is important, since it covers partial completion and provides for the proportionate reduction of liquidated damages where the engineer under clause 48 certifies substantial completion of part of the works.

Sectional completion

Clause 47(2) deals with liquidated damages where the whole of the works is divided into sections that are to be completed within particular times specified in the Appendix to the Form of Tender, and each section is given its own liquidated damages. Clause 47(2)(b) then provides that if the contractor fails to complete a section within the prescribed period of time, liquidated damages become payable. The proviso to the sub-clause allows for proportionate reduction of liquidated damages where part of the section is certified as substantially complete.

The revised wording appears to overcome the difficult provisions of clause 47 in the Fifth Edition where, in relation to works to be completed in sections, the provision was virtually inoperable.

Clause 47(2)(c)) provides that liquidated damages in respect of two or more sections may run concurrently.

'Penalty' clauses

As noted above (see Section 5.2), clause 47(3) is of no legal effect since, as Lord Dunedin remarked in *Dunlop Pneumatic Tyre Co v. New Garage and Motor Co Ltd* (1915), 'the court must find out whether the payment stipulated is in truth a penalty or liquidated damages'. If the court or an arbitrator considered that the figure was so excessive as to constitute a penalty (despite the decision of the Privy Council in *Philips Hong Kong Ltd v. Attorney General*) this clause would not stop the clause from being unenforceable.

Limitations

In the Sixth Edition, provision is made in clause 47(4)(a) for a limit to be placed on liquidated damages if so stated in the Appendix, while clause 47(4)(b) provides that if a sum is omitted from the Appendix as liquidated damages, or any such sum is stated to be 'nil', 'then to that extent damages shall not be payable'. Neither of these provisions appeared in the Fifth Edition, although the proviso to clause 47(1)(a) of that edition provided for a 'lesser sum' than the genuine pre-estimate of the likely loss to be inserted in the appendix as 'the limit of the contractor's liability'.

The only condition precedent to the employer's right to recover liquidated damages is the contractor's failure 'to complete the whole of the works' within the prescribed time (clause 47(1)(b)) or, if appropriate, his failure to so complete a section (clause 47(2)(b)). Accordingly, if the contractor fails to complete by the due date, clause 47(5) empowers the employer to 'deduct' any liquidated damages due 'from any sums due or which become due to the contractor', or to 'require' the contractor to pay the amount due. The engineer does not have to issue any certificate to authorise such deduction. This is quite different from the corresponding provision of the JCT Standard Form of Building Contract.

It is also quite different to the provisions under the Fifth Edition. Under that contract, before the employer could deduct liquidated damages, the engineer had to determine and certify any extension of time to which he considered the contractor was entitled under clause 44(3) (assessment at due date for completion) or clause 44(4) (final determination of extension of time on the issue of a Certificate of Completion). He also had to notify both the employer and the contractor that, in his opinion, the contractor was not entitled to any further extension of time. Only then was the employer entitled to recover liquidated damages from the contractor at the prescribed rate by deduction from the amounts certified by the engineer – clause 47(4), Fifth Edition.

The wording of both Fifth and Sixth Editions makes it clear that it is the employer (and not the engineer) who must effect any deduction. If there is no fund from which to deduct, and the contractor fails to pay any amount that the employer has required him to pay, the employer can recover the amount due by legal process of one sort or another as discussed in Chapter 7.

Reimbursement

Clause 47(5) also provides for the reimbursement (with interest compounded monthly) of liquidated damages which have been deducted where, on a subsequent or final review of the circumstances causing delay, the engineer changes his mind (or an arbitrator decides the issue) and grants an extension or further extension of time. Liquidated damages which have been over-deducted are then reimbursable, together with interest compounded monthly at 2% above the base lending rate of the Bank specified in the Appendix.

Concurrent reasons for delay

Clause 47(6) was introduced in the Sixth Edition to deal with the common situation in which a variation order is issued or a clause 12 situation is encountered during a period of culpable delay. At the time of first publication of the Sixth Edition in 1991 it was thought that the engineer might not have the power to grant an extension of time in such circumstances, and that the new delaying event might have the effect of putting time at large. The 1993 case *Balfour Beatty Building Ltd* v. *Chestermount Properties Ltd*, dealing with a similar problem under the JCT Standard Form, suggests that this concern was ill-founded and that the engineer had power to grant an extension of time (thus preserving the employer's right to liquidated damages). The revised completion date should be calculated by taking the date currently fixed and adding to it the number of days the engineer regarded as fair and reasonable in respect of the consequences of the variation or clause 12 situation: that is, the 'net basis'.

Clause 47(6) deals with the situation by providing that liquidated damages are suspended for the period of delay if, after liquidated damages become payable:

- the engineer issues a variation order under clause 51; *or*
- adverse physical conditions or artificial obstructions within clause 12 are encountered; *or*
- any other situation outside the contractor's control arises; *and*
- the engineer is of opinion that the event results in further delay to the part of the works in respect of which liquidated damages have become payable.

In that event, the engineer must so inform both the contractor and the employer in writing, and the employer's entitlement to damages 'in respect of that part of the Works' is suspended until the engineer notifies both parties in writing that 'the further delay has come to an end'.

The last sentence of the clause is otiose because it is thought that in any event the suspension would not invalidate the employer's entitlement to liquidated damages which had already accrued nor would the happening of the stated events do so: see *Balfour Beatty Building Ltd* v. *Chestermount Properties Ltd* (1993) which supports the view advanced here.

Notice

The March 1998 amendments introduced a further requirement which the employer must satisfy before he can deduct the liquidated damages to which he is entitled from money due to the contractor. The new clause 60(10) reads as follows:

> Where a payment under Clause 60(2) or (4) [interim or final] is to differ from that certified or the Employer is to withhold payment after the final date for payment of a sum due under the Contract the Employer shall notify the Contractor in writing not less than one day before the final date for payment specifying the amount proposed to be withheld and the ground for withholding payment or if there is more than one ground each ground and the amount attributable to it.

This clause was introduced with other amendments to comply with the requirements of the Housing Grants, Construction and

Regeneration Act 1996. If notice is not given of intention to deduct from an interim payment there is a danger that the Contractor might acquire the right to suspend work.

If the employer does not wish to quantify damages for delay in advance, relying instead on his rights to recover such loss as he is able to prove that he suffers as a result of late completion, extensive amendments are required to remove all reference to liquidated damages throughout the contract. It is not sufficient merely to leave the liquidated damages entry in the appendix blank. To do so will mean that he is disqualified from recovering any damages at all, save perhaps from the professional adviser who assisted him to complete the form.

5.3.2   Minor Works Conditions

The Minor Works Conditions provide for liquidated damages in clause 4.6, which reads as follows:

> If by the end of the period or extended period or periods for completion the Works have not reached practical completion the Contractor shall be liable to the Employer in the sum stated in the Appendix as liquidated damages for every week (or *pro rata* for part of a week) during which the Works so remain uncompleted up to the limit stated in the Appendix. Similarly where part or parts of the Works so remain uncompleted the Contractor shall be liable to the Employer in the sum stated in the Appendix reduced in proportion to the value of those parts which have been certified as complete provided that the said limit shall not be reduced.
>
> Provided that if after liquidated damages have become payable in respect of any part of the Works the Engineer issues a variation order under Clause 2.3(a) or an artificial obstruction or physical condition within the meaning of Clause 3.8 is encountered or any other situation outside the Contractor's control arises any of which in the opinion of the Engineer results in further delay to that part of the Works
>
> (a)   the Engineer shall so inform the Contractor and the Employer in writing
>
> and
>
> (b)   the Employer's further entitlement to liquidated damages in respect of that part of the Works shall be suspended until the Engineer notifies the Contractor and the Employer that the further delay has come to an end.
>
> Such suspension shall not invalidate any entitlement to liquidated damages which accrued before the period of delay started to run and

any monies deducted or paid in accordance with this Clause may be retained by the Employer without incurring interest thereon under Clause 7.8.

Clause 4.6 is much more clearly drafted than the corresponding provision in the ICE Conditions, Sixth Edition. It can also operate as a limitation on the contractor's liability if such a provision is stated in the appendix. The Guidance Note recommends that the limit should not exceed 10% of the estimated final contract value and that this should be taken into account when assessing the rate of liquidated damages at the outset. The Note is only for guidance and if a greater limit is set, or no limit at all, the contractor will not be able to rely on it in his defence.

If the works have not reached practical completion by the stipulated time, as extended under clause 4.4, liquidated damages are due at the prescribed rate 'for every week (or *pro rata* for part of a week) during which the Works so remain uncompleted up to the limit stated in the Appendix'. 'Week' is not specially defined and so means a calendar week. Unlike many contracts that use the concept of 'practical completion' this form helpfully provides a definition in clause 4.5(1):

> Practical completion shall occur when the Works reach a state when notwithstanding any defect or outstanding items therein they are taken or are fit to be taken into use or possession by the Employer.

Clause 4.6 goes on to provide for the proportionate reduction of liquidated damages where 'part' of the works is certified as complete under clause 4.5. However, there is no reduction in the overall limit of liquidated damages.

There is then a proviso dealing with new matters that would justify an extension of time when the contractor is in culpable delay and liquidated damages are accruing. The proviso is similar to the arrangement set out in Clause 47(6) of the Sixth Edition discussed above.

There is, however, no requirement that the engineer should have certified late completion; whether or not the works (or part) have reached practical completion is a question of fact. If they have not, the employer is entitled to liquidated damages, and he can recover these by the exercise of his right of set-off against any sums otherwise payable to the contractor (e.g. under certificates) or else by action for debt. As with the Sixth Edition, the March 1998 amendments introduced an obligation at clause 7.10 to give at least one day's written notice of intention to deduct in order to bring the contract into line with the Housing Grants Construction and Regeneration Act 1996.

5.3.3 Interest

No express provision is made for the repayment of liquidated damages if, subsequently, the engineer grants any further extensions of time when he reviews contract progress, so that a refund is due to the contractor. It is clear, however, that in such an event the employer would be bound to repay any sums deducted. There is no provision for interest on any liquidated damages repaid – clause 7.8 deals only with interest on overdue payments or where the engineer fails to certify – but an arbitrator or adjudicator can award interest on an award or decision providing for return of liquidated damages.

It is possible that the contractor might, in legal proceedings, be able to establish a claim to interest as 'special damages', but there is no direct authority. In *Department of the Environment for Northern Ireland* v. *Farrans (Construction) Ltd* (1981), which arose under the JCT Standard Form of Building Contract 1963 edition, the High Court of Justice in Northern Ireland held (on the interpretation of the relevant contract clauses) that the contractor was entitled to recover interest on sums so refunded. Mr Justice Murray held that the arbitrator had power to award damages 'which, if the case is made out, can include interest incurred or lost by the contractor as a foreseeable consequence of the employer's failure to pay on the due date'.

Interest is recoverable as special damages at common law. This is illustrated by *Wadsworth* v. *Lydall* (1981), which was an action for breach of a contract made when a partnership between the parties was dissolved. Under the contract, the plaintiff had to give up possession of a farm, in which he was living, on or before 15 May 1976. In return, the defendant was to pay him £10 000 on that date. On 10 May, the plaintiff agreed to buy a property from a third party and further agreed to pay £10 000 on completion. The defendant did not pay any money until October, and then only £7200. As a result, the plaintiff had to pay the third party interest on the unpaid purchase price of the property, and also incurred costs of raising a mortgage for the balance. The Court of Appeal allowed these two items as special damages.

These principles have been applied by the High Court in the context of a civil engineering contract in *Holbeach Plant Hire Ltd* v. *Anglian Water Authority* (1988). One of the points at issue in that case was whether the contractor was entitled to interest on monies that were received later than they should have been under the terms of the contract, which was based on the ICE Conditions, Fourth Edition. Judge Lewis Hawser QC held that interest on the late payment could be specifically pleaded and proved as a claim for special damages, as distinct from one for general damages.

In principle, there is no reason why a claim for interest on liquidated damages repaid as a result of a further extension of time granted under clause 4.4 should not be pleaded and proved as special damages. Such claims are however seldom successful and the claim for interest is more often made in legal proceedings as a claim for statutory interest, for example under the Arbitration Act 1996.

5.3.4 Liquidated damages preserved by extensions of time

Without an extension of time clause, neither the employer nor the engineer would have any power to extend the contract period. If extra work is added to the contract and completion delayed as a result, the contractor would argue that he was not responsible for the failure to meet the contractual completion date. The strict wording of the liquidated damages provision would then be unworkable and if he also caused a delay by inefficient working the employer would be unable to recover such damages. An extension of time clause is therefore needed so that the employer is able to preserve his rights.

This principle was firmly established in *Percy Bilton Ltd* v. *Greater London Council* (1982). Consequently, all standard form construction contracts contain an extension of time clause, the primary purpose of which is to preserve the employer's right to liquidated damages.

From a practical point of view, the extension of time clause is one of the most important provisions of the contracts. The grant of an extension of time will postpone the moment from which liquidated damages start to accrue. Of itself, the grant of an extension of time does not entitle the contractor to extra payment – additional payments in respect of delay fall to be dealt with under other provisions. It is also important to appreciate that there is no automatic right to an extension of time.

5.4 Extensions of time – the contract provisions

5.5.1 ICE Conditions

Extension of the contractor's time for completion is provided for in clause 44. This reads:

(1) Should the Contractor consider that
   (a) any variation ordered under Clause 51(1) or
   (b) increased quantities referred to in Clause 51(4) or
   (c) any cause of delay referred to in these Conditions or
   (d) exceptional adverse weather conditions or
   (e) any delay impediment prevention or default by the Employer or
   (f) other special circumstances of any kind whatsoever which may occur

be such as to entitle him to an extension of time for the substantial completion of the Works or any Section thereof he shall within 28 days after the cause of any delay has arisen or as soon thereafter as is reasonable deliver to the Engineer full and detailed particulars in justification of the period of extension claimed in order that the claim may be investigated at the time.

(2) (a) The Engineer shall upon receipt of such particulars consider all the circumstances known to him at that time and make an assessment of the delay (if any) that has been suffered by the Contractor as a result of the alleged cause and shall so notify the Contractor in writing.

(b) The Engineer may in the absence of any claim make an assessment of the delay that he considers has been suffered by the Contractor as a result of any of the circumstances listed in sub-clause (1) of this Clause and shall so notify the Contractor in writing.

(3) Should the Engineer consider that the delay suffered fairly entitles the Contractor to an extension of the time for the substantial completion of the Works or any Section thereof such interim extension shall be granted forthwith and be notified to the Contractor in writing. In the event that the Contractor has made a claim for an extension of time but the Engineer does not consider the Contractor entitled to an extension of time he shall so inform the Contractor without delay.

(4) The Engineer shall not later than 14 days after the due date or extended date for completion of the Works or any Section thereof (and whether or not the Contractor shall have made any claim for an extension of time) consider all the circumstances known to him at that time and take action similar to that provided for in sub-clause (3) of this Clause. Should the Engineer consider that the Contractor is not entitled

139

to an extension of time he shall so notify the Employer and the Contractor.

(5)   The Engineer shall within 14 days of the issue of the Certificate of Substantial Completion for the Works or for any Section thereof review all the circumstances of the kind referred to in sub-clause (1) of this Clause and shall finally determine and certify to the Contractor with a copy to the Employer the overall extension of time (if any) to which he considers the Contractor entitled in respect of the Works or the relevant Section. No such final review of the circumstances shall result in a decrease in any extension of time already granted by the Engineer pursuant to sub-clauses (3) or (4) of this Clause.

Grounds for an extension

The grounds which entitle the contractor to an extension of time are very wide. They embrace not only the grounds set out in clause 44(1) itself but also 'any cause of delay referred to *in these Conditions*'. Many delays which are the employer's responsibility are specifically referred to in the conditions, for example late information (clause 7 (4)). The amendments introduced in March 1998 also introduced a further wide category of entitlement – 'any delay impediment prevention or default by the Employer'. The grounds are summarised in Table 5.1. Most are self-explanatory, but two in particular justify some comment:

- *'Other special circumstances of any kind whatsoever'*: Despite the apparent breadth of the wording, this phrase is likely to be given a restrictive interpretation by the courts and so will probably not cover delays caused by the employer not expressly set out as grounds for extension of time: see *Peak Construction (Liverpool) Ltd* v. *McKinney Foundations Ltd* (1970) where it was suggested that an extension of time clause should be construed against the employer in whose favour it operates by application of the *contra proferentem* rule. This possible problem for employers has been addressed in the March 1998 amendments, which introduced the current clause 44(1)(e).
- *'Engineer's failure to issue any necessary drawings or instructions'*: The engineer is only required to issue drawings etc 'at a time reasonable in all the circumstances'. He is not necessarily required to issue his drawings in accordance with the contractor's contract programme, which may be based on early completion: *Glenlion Construction Ltd* v. *The Guinness Trust Ltd* (1987). Moreover the drawings or other information must have

Table 5.1 Grounds for extension of time

GROUND FOR EXTENSION OF TIME	CLAUSE
Variations ordered under clause 51(1)	44(1)
Increased quantities referred to in clause 51(4)	44(1)
Exceptional adverse weather conditions	44(1)
Any delay impediment prevention or default by the employer	44(1)
Other special circumstances of any kind whatsoever	44(1)
Engineer's failure to issue any necessary drawings or instructions at reasonable times	7(4)
Adverse physical conditions or artificial obstructions	12(2)
Instructions under clauses 5 or 13(1)	13(3)
Engineer's delay in giving consent to contractor's proposed methods of construction etc.	14(8)
Variations involving street works etc.	27(6)
Facilities for other contractors	31(2)
Suspension of progress of the works	40(1)
Failure to give sufficient possession of site	42(3)

been requested by the contractor and be considered necessary by the engineer before delay in issue can justify an extension of time.

Nominated sub-contractors

It is sometimes suggested by contractors, perhaps because of their experience under JCT building contracts, that there is a right to an extension of time for delay caused by nominated sub-contractors. This is not so in the case of simple delay caused by a nominated sub-contractor's breach of sub-contract or other default. This is not a fault of the employer: *Percy Bilton Ltd* v. *Greater London Council* (1982). Clause 59(3) states specifically that the main contractor is responsible for the nominated sub-contractor, and nothing elsewhere in the contract affects this position. (See Chapter 3, section 3.22.)

If a nominated sub-contractor fails, the immediate financial consequences fall upon the main contractor, although the employer will be liable for any delay in making a fresh nomination: see the building contract case of *Fairclough Building Ltd* v. *Rhuddlan Borough Council* (1985) and clause 59(2) as to the action to be taken by the engineer on determination of a nominated sub-contract.

Actual/planned progress

The wording of clause 44 is extremely important. The mere happening of a delaying event confers no entitlement to an extension of the contract period. The test to be applied is whether the event is 'such as to entitle the Contractor to an extension of time' for completion of the works, or where appropriate for any relevant section of them, and the contractor's actual progress must in fact be delayed. The position is the same in this respect under both Fifth and Sixth Editions, although in the former the reference was to events 'such as may fairly entitle the Contractor to an extension of time'.

The contractor's planned or programmed progress is irrelevant. His obligation under clause 43 is to complete 'within' the stated time. He may complete earlier if he so wishes, and indeed he may have priced internally to do so, but unless there is delay beyond the time stated in the contract itself there is no entitlement to an extension of time. If the contractor is able to finish the work in time, in spite of the delay, he has no right to an extension of time.

The engineer's acceptance of the programme showing the sequence and time-scale in which the contractor proposes to carry out the works under clause 14(1) does not affect the basic contractual obligations or responsibilities of the parties to the contract. The acceptance is merely of evidentiary value (see *Glenlion Construction Ltd* v. *The Guinness Trust Ltd* (1987)).

Concurrent causes of delay

Considerable difficulties arise when evaluating extensions of time, particularly where there are concurrent causes of delay, where one delay is the contractor's own fault, and the other would entitle him to an extension of time. The crucial question here appears to be what is the dominant cause of the delay. If, for example, the contractor is late through his own fault, and concurrently the engineer is late in providing him with necessary information, the contractor is only entitled to an extension of time if in fact he could have acted upon the information. There is no mechanism under the contract for allocating the causes of delay between different heads, and any other interpretation would produce absurdity.

The procedure

Clause 44 sets out a three-stage procedure for the grant of extensions of time in respect of qualifying delays. The procedure is designed to prevent time becoming 'at large', and has great merit. The requirement of notice from the contractor is not a condition precedent to the operation of the clause, since at the due date for substantial completion the engineer is under an independent duty to consider whether any extension of time should be granted (clause 44(4)). Indeed if it was a condition precedent and no extension could be given without application from the contractor, any instruction requiring additional work could have the effect of putting time at large and the employer would lose his right to claim liquidated damages for delay. If the contractor fails to submit a claim for extension of time under clause 44(1) when a cause of delay arises, this merely postpones his entitlement to an extension of time because, as a result of clause 44(5), the engineer is bound to consider whether or not any extension of time is merited even if the contractor has made no claim at all.

Clause 44(1) requires the contractor to deliver to the engineer full and detailed particulars *in justification of the period of extension claimed* within 28 days of the cause of delay arising 'or as soon thereafter as is reasonable'. The time runs from when the delay occurred and in practice it may be some time before the 'full and detailed particulars' can be prepared. The purpose of requiring the particulars of claim is so that the claim 'may be investigated at the time'. It is in the contractor's interest to submit his claim to the engineer as quickly as possible because the engineer is then obliged to consider it and assess any extension of time which it merits. The wording of clause 44(1) in the Sixth Edition makes it plain that the contractor must make a claim for a specific time period, although the form of the particulars or the detail required is not specified. The contractor should give the engineer as much information as he can to assist the engineer in performing his duty.

The engineer's duty as regards interim assessments of time is set out in clause 44(2). 'Upon receipt' of the contractor's particulars of claim the engineer must consider 'all the circumstances known to him at that time' and assess any delay that has been suffered by the contractor as a result of the alleged cause of delay. He must then notify the contractor in writing. This assessment should be made promptly upon the receipt of the contractor's claim. In the absence of a claim from the contractor the engineer may make an assessment in respect of any delay that he considers the contractor has suffered as a result of any of the circumstances referred to in clause 44(1).

Once he has made an assessment of delay under clause 44(2), and he considers that it *fairly* entitles the contractor to an extension of time, the engineer must grant any extension that is justified by written notice to the contractor. If the contractor has made a claim under clause 44(1) (but not otherwise) and the engineer considers that the contractor has no entitlement to any extension of time, he is bound to inform the contractor accordingly.

In the nature of things, an interim assessment is likely to be the minimum to which the engineer considers the contractor to be entitled because he has no power to decrease any extension of time once granted, even on the final determination of extension of time under clause 44(5). He can, however, give a further extension by his final determination. If the contractor wishes to make an immediate challenge to the engineer's interim assessment he can do so through the dispute resolution processes described in Chapter 7. The engineer may not be acting in the employer's best interests if he is unduly conservative in dealing with extensions of time. If he fails to grant a justified extension the employer may well be found to be in breach of contract. Acceleration costs incurred by the contractor in efforts to overcome the relevant delay might then be treated as damages for that breach, and recovered through a common law claim.

It is clear from the revised wording of clause 44(3) that an extension of time need only be granted if the engineer is of the opinion that it is actually needed for the completion of the works (or a section).

In making his assessment of the delay, the engineer is required to consider all the circumstances known to him at the time and not just those to which the contractor has referred. The contractor's own performance and omissions from the works may both be relevant.

Not later than 14 days after the due date or extended date for completion of the works or any section, whether or not the works are complete, the Engineer is required to assess any extension of time due, even if the contractor has made no claim for an extension of time, and to notify both the contractor and the employer if he considers that the contractor is not entitled to any extension of time. Once again the engineer has to consider all the circumstances known to him at the time. The notification to the employer is so that he may begin to deduct liquidated damages if appropriate. Under the Fifth Edition no time limit was imposed but the engineer's duty had to be performed 'at or as soon as possible after' the due or extended date for completion.

If in the exercise of his duty under clause 44(4) the engineer grants an extension of time, he thereby creates a new 'extended date for completion'. He must again make an assessment under clause 44(4) at that new date. At that point, he must reconsider matters in light of any delaying events which have occurred subsequently.

For example, the engineer has granted an interim extension of time as a result of a variation instruction, and grants a further extension of time at the revised date for completion. A strike then occurs, and it takes the contractor longer to complete the work as a result. The engineer is bound in those circumstances to grant a further extension of time under clause 44(4).

The engineer is under a duty to review the situation yet again within 14 days of the issue of a certificate of substantial completion for a section or the whole of the works, even if the contractor has made no claim at all for extension of time in respect of the particular section or the works as a whole (clause 44(5)). The engineer cannot at this stage reduce an extension given previously. This is is the final opportunity for the engineer to determine any overall extension of time to which the contractor is entitled unless and until the disputes procedures are invoked.

5.5.2 Minor Works Conditions

Clause 4.4 provides as follows:

If the progress of the Works or any part thereof shall be delayed for any of the following reasons:
(a)   an instruction given under Clause 2.3(a) (c) or (d)
(b)   an instruction given under Clause 2.3(b) where the test or investigation fails to disclose non-compliance with the Contract
(c)   encountering an obstruction or condition falling within Clause 3.8 and/or an instruction given under Clause 2.3(e)
(d)   delay in receipt by the Contractor of necessary instructions drawings or other information
(e)   failure by the Employer to give adequate access to the Works or possession of land required to perform the Works
(f)   delay in receipt by the Contractor of materials to be provided by the Employer under the Contract
(g)   exceptional adverse weather
(h)   any delay impediment prevention or default by the Employer
(j)   other special circumstances of any kind whatsoever outside the control of the Contractor
then provided that the Contractor has taken all reasonable steps to avoid or minimize the delay the Engineer shall upon a written request by the

Contractor promptly by notice in writing grant such extension of the period for completion of the whole or part of the Works as may in his opinion be reasonable. The extended period or periods for completion shall be subject to regular review provided that no such review shall result in a decrease in any extension of time already granted by the Engineer.

Clause 4.4 is less complicated in its operation than clause 44 of the ICE Conditions; it is also better drafted. Nine grounds which may give rise to an extension of time are specified. These are:

- Instructions requiring a variation, suspension of the whole or part of the works, or requiring any change in the intended sequence of the works.
- Instructions requiring testing or investigation where the test or investigation fails to disclose non-compliance with the contract.
- Adverse physical conditions and obstructions and/or consequent engineer's instructions.
- Delay in receipt by the contractor of necessary drawings, instructions or other information from the engineer.
- Failure by the employer to give adequate access to the works or possession of land required to perform them.
- Delayed receipt of materials to be provided to the contractor by the employer under the contract.
- Exceptional adverse weather.
- Any delay, impediment, prevention or default by the employer.
- Other special circumstances of any kind whatsoever outside the control of the contractor.

Although the wording is not identical to the equivalent clauses of the ICE Conditions, similar considerations apply.

The test to be applied by the engineer is whether the actual progress of the works or part of the works is delayed by one or more of the specified events. Once again, it is the contractor's actual progress that is important, and the contractor's programme has no more than evidentiary value.

Unlike the equivalent clause in the ICE Conditions, the Minor Works Form imposes a proviso that the contractor must have taken all reasonable steps to avoid or minimise the delay to progress. What steps are 'reasonable' will depend on the circumstances, but it is thought that the contractor is not required to spend substantial sums of money. In some cases meeting this obligation may require the contractor to reprogramme the works to avoid or minimise the delay, if necessary taking into account and using any 'float' built into his programme. What is necessary to meet the obligation depends on the circumstances.

Whilst the clause refers to a written request by the contractor such a notice is not a condition precedent to the exercise of the duty to grant an extension of the period for completion. The contractor's failure to make a written request for an extension of time is however a factor that the engineer may take into account in assessing what is a reasonable extension. No particular form of request is specified, except that it must be in writing, but it is in the contractor's interest to provide the engineer with as much information as possible.

If the engineer becomes aware that progress of the works, or any part of them, is delayed because of an event falling within clause 4.4, he should grant an appropriate extension of time whether or not the contractor has made a written request to him. He should do this before the current date for completion of the works (or the appropriate part of them) has passed – certainly if the delay is caused by employer default, and certainly no later than the issue of the final certificate for payment when he becomes *functus officio*.

On receipt of a written request for extension of time from the contractor, the engineer should consider it, assess the extension of time that it merits, and notify the contractor in writing of the extension granted. It is intended that the extension be granted promptly upon receipt of the claim. The common practice of postponing consideration of extensions of time until the end of the contract is not in accordance with this clause.

Little is said about the circumstances that the engineer must take into account in assessing what extensions to grant. There are no defined criteria except that the whole or part of the works is delayed and the contractor must have taken all reasonable steps to avoid or minimise the delay.

The extension to be granted is one that may be reasonable in the opinion of the particular engineer. Within limits this is a personal decision, and different engineers might reach slightly different conclusions on the same facts. However, the engineer is under a duty to act fairly to both the contractor and the employer, and the reasonableness of the engineer's opinion can be challenged under the disputes procedure. Clause 11.7 confers on the arbitrator 'full power to open up review and revise any decision instruction certificate or valuation of the engineer', and in effect to substitute his own opinion for that of the engineer. In the light of the decision in

Beaufort Developments (NI) Ltd v. *Gilbert Ash NI Ltd* (1998), the court has a similar power in litigation.

The final sentence of clause 4.4 requires the engineer to review 'the extended period or periods for completion' regularly, but his 'regular review' cannot decrease any extensions already granted. In other words, the contractor will not find himself liable for liquidated damages because an extension of time already granted is reduced. The contractor's liability for liquidated damages (clause 4.6) only arises if the works (or part thereof) have not reached practical completion 'by the end of the period *or extended period or periods* for completion'.

CHAPTER SIX
COMMON LAW CLAIMS

6.1 Introduction

In the preceding chapters discussion has centred on claims that are permitted under specific provisions of the contracts and which can be the subject of certificates or determinations of the engineer. Such contractual claims are not exhaustive of the contractor's rights and remedies. He may bring additional or alternative claims (sometimes based on the same facts) at common law. There is no provision in either the ICE Conditions or the Minor Works Contract that excludes or limits such a claim.

The engineer cannot deal with such claims – unless he is expressly authorised by the employer to do so and the contractor agrees to his settling the claim. He may be asked to give a decision on such a common law claim under clause 66 of the ICE Conditions as a precursor to arbitration or other proceedings, but he will not be able to certify a common law entitlement under the contract. Indeed, in the absence of agreement, common law claims must in principle be pursued in arbitration, litigation or adjudication. It is often appropriate in formal legal proceedings to include alternative claims under the contract and at common law.

Most common law claims are based on alleged breach of implied terms in the contract. There can be claims for breach of express terms as well, and also claims in tort and for misrepresentation.

6.2 Claims for breach of implied terms

Neither the ICE Conditions nor the Minor Works form are self-sufficient documents containing all the contractual obligations of the parties. They – and other standard form contracts – are firmly fixed in the mainstream of the general law.

Both sets of printed conditions contain their own definition of the 'contract'. These definitions suggest that the 'contract' as defined is completely self-contained. So, in the ICE Conditions, clause 1(1)(e) defines the 'Contract' as meaning:

> [The] Conditions of Contract Specification Drawings Bill of Quantities the Tender the written acceptance thereof and the Contract Agreement (if completed).

The corresponding provision in the Minor Works Conditions is clause 1.2. This says that the 'Contract' means 'the Agreement if any together with these Conditions of Contract the Appendix and other items listed in the Contract Schedule'. (The omission of any punctuation in these quotations is that of the draftsmen.)

These definitions suggest to non-lawyers that everything is contained within the 'Contract' as so defined; but the printed conditions and other documents referred to contain only the *express terms* agreed by the parties. Indeed, in some cases those documents may not contain all the express terms because it is perfectly possible for the parties to agree other terms orally, though this would be unusual in the context of a civil engineering contract.

There is also the possibility that in some cases the courts (or an arbitrator) will write terms into the contract in order to make it commercially effective. These are called *implied terms,* and breach of an implied term can also give rise to a claim for damages. Clause 49(3) of the ICE Conditions, Fifth Edition, referred explicitly to the existence of implied terms by speaking of 'any obligation expressed *or implied* on the Contractor's part under the Contract', and although these words have disappeared in the Sixth Edition, it is clear that terms may still be implied.

The subject is a complex one, since terms may be implied in a number of ways.

6.2.1   By custom (commercial or local)

This is of little importance in the civil engineering field, but see, for example *Produce Brokers Co Ltd* v. *Olympia Oil and Cake Co Ltd* (1916).

6.2.2   By statute

In the nineteenth century and earlier, the courts imposed implied terms on the parties to all contracts of certain types, such as sale of goods and bills of exchange, by way of judicial legislation. Such implied terms have been consolidated by statute and, as Judge Peter Bowsher QC remarked in *Barratt Southampton Ltd* v. *Fairclough Building Ltd* (1988):

> 'These terms now have the status of standard terms of contract upon which everyone is deemed to contract unless express terms are agreed to the contrary. There is little to be learned from them when considering what terms should be implied into a modern commercial contract falling outside those old cases or the consolidating statutes into which they have been incorporated.'

A straightforward instance of this type of term is to be found in the field of sale of goods. Where there is a contract for the sale of goods, the Sale of Goods Act 1979 implies various terms into the contract, for example that the goods will be reasonably fit for any purpose made known to the seller.

A similar term will be implied in contracts for the supply of a service or contracts for work and materials by section 4 of the Supply of Goods and Services Act 1982. This affects the civil engineering world since the engineer is a person who 'agrees to carry out a service'. As a result, a consultant engineer acting in the course of business is under a statutorily implied obligation to carry out his work, for example as a designer, with reasonable care and skill. Other terms are also implied.

The Housing Grants, Construction and Regeneration Act 1996 introduces terms into contracts in a slightly different manner. It requires all construction contracts, as defined by the Act, to include terms which deal with specific aspects of payment entitlements and also provide the right to refer disputes at any time to adjudication. If the contract does not contain terms which deal with these aspects, the statutory Scheme for Construction Contracts will be deemed to apply.

6.2.3 By the courts

Lord Justice Bowen in his judgment in *The Moorcock* (1889) said this:

> 'An implied warranty ... is in all cases founded on the presumed intention of the parties, and upon reason. The implication which the law draws from what must obviously have been the intention

of the parties, the law draws with the object of giving efficacy to the transaction... In business transactions such as this, what the law desired to effect by the implication is to give such business efficacy to the transaction as must have been intended by both parties who are business men.'

In 1939 this was expanded by Lord Justice MacKinnon in *Shirlaw* v. *Southern Foundries (1926) Ltd*:

'Prima facie that which in any contract is left to be implied and need not be expressed is something so obvious that it goes without saying; so that, if, while the parties were making their bargain, an officious bystander were to suggest some express provision for it in their agreement, they would testily suppress him with a common "Oh, of course!".'

More recently it has been said (by Lord Simon in *BP Refinery Ltd* v. *Shire of Hastings* (1977) that for a term to be implied

'the following conditions (which may overlap) must be satisfied:
(1)   it must be reasonable and equitable;
(2)   it must be necessary to give business efficacy to the contract so that no terms will be implied if the contract is effective without it;
(3)   it must be so obvious that "it goes without saying";
(4)   it must be capable of clear expression;
(5)   it must not contradict any express term of contract.'

Examples of the court's approach – clear express terms

Thus, where the express terms of the contract are clear and unambiguous, the courts will not imply a term simply to extricate a party from difficulties: *Trollope & Colls Ltd* v. *North-West Metropolitan Regional Hospital Board* (1973). In that case, building works were to be carried out in three phases. There was a separate contract sum and set of conditions for each phase, all of which were in a common standard form. The Phase III commencement date was to be fixed by reference to the completion date of Phase I, although there was a specified completion date for Phase III. There was a delay to Phase I and the contractor was granted an extension of time of 57 weeks. This allowable delay effectively reduced the period for completing Phase III from 30 to 16 months.

The contractor claimed that a term extending the completion date of Phase III should be implied. This argument was rejected by the House of Lords because the contract was clear and unambiguous. The Phase III completion date was clearly stated.

Lord Pearson put the matter clearly:

'The court does not make a contract for the parties. The court will not even improve the contract which the parties have made for themselves, however desirable the improvements might be. The court's function is to interpret and apply the contract which the parties have made for themselves. If the express terms are perfectly clear and free from ambiguity, there is no choice to be made between possible meanings: the clear terms must be applied even if the court thinks some other terms would have been more suitable. An unexpressed term can be implied if and only if the court finds that the parties must have intended that term to form part of their contract; it is not enough for the court to find that such a term would have been adopted by the parties as reasonable men if it had been suggested to them: it must have been a term that went without saying, a term *necessary* to give business efficacy to the contract, a term which, though tacit, formed part of the contract which the parties made for themselves.'

Replacement engineer

It is necessary under the ICE Conditions for there to be a named engineer in order to make the contract effective. Clause 1(1)(c) defines 'the Engineer' as meaning:

the person firm or company appointed by the Employer to act as Engineer for the purposes of the contract and named in the Appendix to the Form of Tender or such other person, firm or company so appointed from time to time by the Employer and notified in writing as such to the Contractor.

If, for any reason, the named engineer dropped out of the picture – for instance, being a named individual he died or resigned his appointment – it would be an implied term of the contract that the employer would appoint a successor even if there were no express term to that effect.

This was the decision in the case of an architect under a JCT 63 building contract in *Croudace Ltd* v. *London Borough of Lambeth* (1986) where the council failed to appoint a successor architect when the appointed architect retired. Both at first instance and on appeal it was held that there was an implied obligation on the employer to nominate a successor architect if the person designated as 'the Architect' under the contract retired or resigned his appointment. The Court of Appeal held that the council's acts and omissions, including but not limited to their failure to appoint a successor architect, amounted to a failure by them to take the necessary steps to settle the contractor's money claim made under the express provisions of the contract, and that as such amounted to a breach of contract. An interim payment of £100 000 was ordered in respect of the claim despite the absence of an architect's certificate.

The same principle clearly applies to the ICE Conditions. If the named engineer drops out of the picture, the employer would be bound to appoint his successor within a reasonable tine. The point does not arise under the Minor Works form because clause 2.1 makes it an express term of the contract that the employer shall appoint a successor.

Practical phrasing

In *Whittal Builders Co Ltd* v. *Chester-le-Street District Council* (1987) there was a contract for the modernisation of 108 dwellings. Before the formal contract was executed, discussions took place between the parties and it was agreed that the contractors would complete the work in 18 months, provided the council gave them possession of at least 18 houses at any one time.

The major part of the work was the building of an extension at the rear of the houses. In all but six cases, the extension was a single building extending out to the rear of two houses with an internal party wall, designed to be built as one building only. This was made clear in both the contract drawings and specification. The completion date was stated, and after it was added 'provided that 18 dwellings be available for the contractor to work in at any one time'.

The High Court upheld the contractor's argument that it was entitled not only to possession of 18 houses at any one time but also, on the basis of an implied term, was entitled to the possession of the houses in pairs. This was necessary to make the contract work. The Court applied the test approved by the House of Lords in *Trollope & Colls Ltd* v. *North-West Metropolitan Regional Hospital Board* (1973), and also referred to the opinion of Lord Wright in *Luxor (Eastbourne) Ltd* v. *Cooper* (1941) where he said:

'It is well recognised that there are cases where obviously some term must be implied if the intention of the parties is not to be defeated, some term of which it can be predicated that "it goes without saying", some term not expressed but necessary to give to the transaction business efficacy as the parties must have intended. This does not mean that the court can embark on a reconstruction of the agreement on equitable principles, or on a view of what the parties should, in the opinion of the court, reasonably have contemplated. The implication must arise inevitably to give effect to the intention of the parties.'

Hindrance and prevention

The majority of contractor's claims will be based on alleged breach of implied terms by the employer that he will co-operate with the contractor in securing the satisfactory performance of the contract and that he will not hinder or prevent the contractor from achieving the desired outcome. It is sometimes argued that such terms are rendered unnecessary by the web of rights and obligations that is contained in a standard form of contract as comprehensive as the ICE Conditions, or that to imply such terms would upset the balance of allocation of risks achieved by the standard negotiated conditions. This argument is unsound, because both the ICE Conditions and the Minor Works Conditions clearly state where the employer (through the agency of the engineer) is permitted to hinder the contractor's progress or cause him to incur additional cost or expense, and in those cases the appropriate remedy is provided under the contract. The argument has been specifically rejected in Hong Kong in relation to a similarly complex contract: *Jardine Engineering Co Ltd* v. *Shumitzu Corporation* (1993).

Neither the ICE Conditions nor, *a fortiori*, the Minor Works form, deal exhaustively with every conceivable act by an employer that might hinder the contractor. Those acts must therefore be covered by an implied term. It is also an implied term under both contracts that the employer will do all that is reasonably necessary on his part to bring about completion of the contract, that is, to co-operate.

The implication of terms of this sort has been considered by the English courts in connection with the JCT Standard Form of Building Contract, which is just as comprehensive as the ICE Conditions, as well as in other jurisdictions on different forms of construction contract. For example, in *Perini Corporation* v. *Commonwealth of Australia* (1969), terms were implied into the contract that:

- the employer should not interfere with the proper performance by the engineer of the duties imposed upon him by the contract; and
- the employer was bound to ensure that the engineer performed the various duties imposed upon him by the contract.

Mr Justice Macfarlane was emphatic that both the negative and positive terms should be implied.

The practical position is well illustrated by *London Borough of Merton* v. *Stanley Hugh Leach Ltd* (1985) where the High Court dealt with the contractor's argument that terms were to be implied in a JCT 63 building contract. The works, the construction of 287 'dwellings', were 101 weeks late in reaching completion. Leach argued that the delays were largely the fault of the architect. His failings, they said, were effectively breaches of implied terms by the employer. The terms on which they relied included:

- That the employer would not hinder or prevent the contractor from carrying out its obligations in accordance with the terms of the contract and from executing the works in a regular and orderly manner.
- That the employer would take all steps reasonably necessary to enable the contractor to discharge its contractual obligations and to execute the works in a regular and orderly manner.

These implied terms are two faces of the same coin. As regards the first or negative term in the *Leach* case, Mr Justice Vinelott considered and applied a statement of Lord Justice Vaughan Williams in *Barque Quilpué Ltd* v. *Brown* (1904):

'There is an implied contract by each party that he will not do anything to prevent the other party from performing a contract or to delay him in performing it. I agree that generally such a term is by law imported into every contract.'

This undertaking is not qualified by the express terms of either the ICE Conditions or the Minor Works form and, as Mr Justice Vinelott said:

'[The] general duty remains save so far as qualified. It is difficult to conceive of a case in which this duty could be wholly excluded.'

The general duty is not 'wholly excluded' under either set of conditions and the contractor may claim at common law for breach of this implied term if he suffers damage as a result.

The positive term is equally applicable to civil engineering contracts. It is an implied term of such contracts that both parties will do whatever is necessary in order to enable the contract to be carried out. Mr Justice Vinelott cited the views of Lord Blackburn in *Mackay* v. *Dick* (1881), which are of general application to both building and civil engineering contracts:

'I think I may safely say as a general rule that where in a written contract it appears that both parties have agreed that something shall be done which cannot effectually be done unless both concur in doing it, the construction of the contract is that each agrees to do all that is necessary to be done on his part for the carrying out of that thing, though there may be no express words to that effect.'

Provision of information

Civil engineering contracts require close co-operation between the contractor and the engineer. It would be pointless to attempt to give an exhaustive list of the many examples of co-operation which the engineer is required to give the contractor.

A single example illustrates the situation. Clause 7(1) of the ICE Conditions, Sixth Edition, requires the engineer to supply the contractor with such further or modified drawings as may be necessary to enable the contractor to complete the works. In *London Borough of Merton* v. *Stanley Hugh Leach Ltd* (1985) the court held that a third implied term that the architect would provide correct information to the contractor was merely a particular application of the general positive term requiring co-operation.

Although in one other English case *(Holland Hannen & Cubitt (Northern) Ltd* v. *Welsh Health Technical Services Organisation* (1981)) similar but more widely expressed terms to those held to apply by Mr Justice Vinelott were conceded, there are obviously limitations. This the judge recognised, since he declined to express any view on whether there was another implied term requiring the architect

'to administer the contract in an efficient and proper manner and in accordance with the practice and procedure normally followed by architects administering a substantial building contract.'

The first part of this term was another particular application of the employer's contractual undertaking that there would be a reasonably competent and skilful architect. What steps the architect had to take in order to discharge his duty was a matter for the arbitrator. In principle, however, it seems that such a term is to be implied in construction contracts of a substantial size.

There is no room for implied terms if those expressly agreed rule them out. This is neatly illustrated in a sub-contract situation by *Martin Grant & Co Ltd* v. *Sir Lindsay Parkinson & Co Ltd* (1984), a decision of the Court of Appeal.

Grant were formwork sub-contractors to Parkinson. The main contracts were in a standard form, but the sub-contracts were not and contained no provisions corresponding to those in the head contracts relating to extensions of time and contractual claims for loss and expense. There were substantial delays in the performance of the main contracts. As a result, Grant found themselves carrying out the work several years later than they had anticipated, and they incurred heavy losses. Grant claimed damages for breach of contract against Parkinson and contended, *inter alia*, that a term was to be implied into the sub-contracts:

'That (a) [the main contractor] would make sufficient work available to the [sub-contractors] to enable them to maintain reasonable progress and to execute their work in an efficient and economic manner; and (b) [the main contractor] should not hinder or prevent [the sub-contractors] in the execution of the sub-contract works.'

Both the trial judge and the Court of Appeal decided that the express terms of the sub-contracts left no room for such implied terms. By clause 2 of the sub-contracts, Grant had undertaken to 'do and perform all the obligations imposed upon or undertaken by the main contractor under the [head contract] ... *at such time or times and in such manner as the [main] contractor shall direct or require'*.

Clause 3 of each sub-contract required the sub-contractors to 'proceed with the [sub-contract] works expeditiously and punctually to the requirements of the [main] contractor ... at such time or times as the [main] contractor shall require having regard to the requirements of the [main] contractor in reference to the progress or condition of the main works and shall complete the whole of the [sub-contract] works to the satisfaction of the [main] contractor'.

These clauses meant that there was a clear risk for Grant that the main and sub-contracts might go on much longer than was originally intended, as turned out to be the case. It was argued for Grant that the terms should be implied because of the relationship between the parties. Since the law imposes on main and sub-contractors an obligation to co-operate with each other to ensure successful performance, it was said that the very relationship between them required the express contractual terms to be interpreted so as to ensure that co-operation was not vitiated.

This view was decisively rejected by the Court of Appeal. As Lord Justice Lawton put it, 'the degree of co-operation will depend ... upon the express terms of any contract which the parties may have made'.

The basic implied terms discussed earlier, however, are not displaced by either the ICE Conditions or Minor Works Conditions, and in appropriate circumstances the employer is liable for breach of them at common law. Those terms mentioned are not exhaustive.

Possession and supply of materials

In the absence of express terms dealing with the time and extent of possession, for example, it would be an implied term of a civil engineering contract that the employer gave possession of the site to the contractor in sufficient time to enable him to complete the works by the due date: *Freeman & Son* v. *Hensler* (1900). The degree of possession to be given depends on the circumstances of the case, but the Canadian Federal Court of Appeal has held *(The Queen* v. *Walter Cabott Construction Ltd* (1975)) that it is fundamental in any construction contract that workspace be provided unimpeded by others for whom the employer is responsible or over whom he has control.

There are many other possibilities. For example in *Thomas Bates & Son Ltd* v. *Thurrock Borough Council* (1975) components were to be provided by the employer. The Court of Appeal held that there was an implied term in the contract that the components would be supplied in time to enable the contractor to carry out the works expeditiously and in accordance with his planned progress. The employer was there held liable for his supplier's default.

Reasonable and necessary

Limitations on implied terms were considered in *J & J Fee Ltd* v. *The Express Lift Company* (1993). Express argued that there must be an implied term in their sub-contract that Fee would provide them with correct and comprehensive information at appropriate times to enable them to fulfil their obligations. Fee were prepared to accept that but argued that the term should be subject to a written request for information being made. Express were successful because the term for which they argued was reasonable and necessary. The Official Referee decided the requirement for a written request was not 'necessary', and therefore Fee's argument failed.

6.3   Other common law claims

6.3.1   Tort

The conditions of contract are not a complete code of obligations between the parties. The law of tort must also be considered. (A tort is a civil wrong other than a breach of contract or a breach of trust or other merely equitable obligation and which gives rise to an action for unliquidated damages at common law: Sir John Salmond.)

In some cases a contractor may found a claim based on breach of duty arising under the law of tort, but recent important developments in the law of negligence have severely curtailed the possibility of a remedy in tort where there has been a contractual assumption of responsibility: *Greater Nottingham Co-operative Society* v. *Cementation Piling & Foundations Ltd* (1988). The judicial trend is now to restrict negligence liability.

Thus, in *Hiron* v. *Pynford Services Ltd* (1991), Judge John Newey QC, having considered all the authorities in a very careful judgment, discussed the possibility of there being liability in both contract and tort. In his view 'the mere existence of a contract does not preclude liability in tort'. The correct approach

> 'is that where there is a contract between [the parties] the contract should be treated as a very important consideration in deciding whether there was sufficient proximity between them and whether it is just and reasonable that the defendant should owe a duty in tort.'

On the facts of *Hiron*, however, the judge held that it was not 'just and reasonable' to find liability in tort where there was a contract. 'If that had been the intention of the parties', he said, 'they could have provided for it in their contract'.

On the other hand, in *Barclays Bank plc* v. *Fairclough Building Ltd* (1993) the existence of a contract was held not to preclude a duty of care in tort. See further Section 6.3.2 below.

6.3.2 Misrepresentation, misinformation and 'reliance'

A misrepresentation is a false statement of fact made during pre-contractual negotiations and which is one of the inducing causes of the contract. Misrepresentations which do not become part of the contract may give rise to liability both at common law and under the Misrepresentation Act 1967. There is also the possibility of liability for negligent misstatement: *Hedley Byrne & Co* v. *Heller & Partners Ltd* (1963); *Caparo Industries plc* v. *Dickman and Others* (1990).

Claims for misrepresentation may be raised in arbitration or litigation, as illustrated by the building contract case of *Holland, Hannen & Cubitt (Northern) Ltd* v. *Welsh Health Technical Services Organisation* (1980) which ended in a settlement of £1.6 million in favour of the contractors. Amongst other things, the contractors claimed that they were 'induced into the contract as the result of misrepresentations and warranties' contained in:

- An item headed 'sequence of operations' in the preliminaries section of the bills of quantities which allegedly induced them to tender on a particular basis.
- Statements made in pre-contractual letters from the architects and at pre-contract meetings.

Statements in the bills or other documentation provided by the employer or the engineer which amount to misrepresentations may give rise to claims. The contractor may have a claim against the employer for misrepresentation about site conditions and other risks made during the negotiations leading to the contract. This is so despite the existence of clause 11 of the ICE Conditions which provides:

(1) The Employer shall be deemed to have made available to the Contractor before the submission of his Tender all information on
    (a)   the nature of the ground and subsoil including hydrological conditions and
    (b)   pipes and cables in on or over the ground
obtained by or on behalf of the Employer from investigations undertaken relevant to the Works.

The Contractor shall be responsible for the interpretation of all such information for the purposes of constructing the Works and for any design which is the Contractor's responsibility under the Contract.

(2)   The Contractor shall be deemed to have inspected and examined the Site and its surroundings and information available in connection therewith and to have satisfied himself so far as is practicable and reasonable before submitting his Tender as to

   (a)   the form and nature thereof including the ground and sub-soil
   (b)   the extent and nature of work and materials necessary for constructing and completing the Works and
   (c)   the means of communication with and access to the Site and the accommodation he may require

and in general to have obtained for himself all necessary information as to risks contingencies and all other circumstances which may influence or affect his Tender.

(3)   The Contractor shall be deemed to have

   (a)   based his Tender on the information made available by the Employer and on his own inspection and examination all as aforementioned and
   (b)   satisfied himself before submitting his Tender as to the correctness and sufficiency of the rates and prices stated by him in the Bill of Quantities which shall (unless otherwise provided in the Contract) cover all his obligations under the Contract.

The situation is illustrated by *Morrison-Knudsen International Co Inc v. Commonwealth of Australia* (1972), where the contract contained a clause which said:

'The contractor acknowledges that he has satisfied himself as to the nature and location of the work, the general and local conditions, including ... the structure and condition of the ground ... Any failure by the contractor to acquaint himself with the available information will not release him from estimating properly any difficulty or the cost of successfully performing the work. The [employer] assumes no responsibility for any conclusions or interpretations made by the contractor on the basis of information made available by the [employer].'

   At the pre-tender stage, the employer provided the contractor with basic information about the site soil conditions in a document called 'Engineering Site Information', which also contained a disclaimer of liability. The contractor claimed that the information provided 'was false, inaccurate and misleading [and] the clays at the site, contrary to that information, contained large quantities of cobbles'.

On a preliminary issue, the High Court of Australia held that the documents did not disclose that the contractor had no cause of action and Chief Justice Barwick said:

'The basic information in the site information document appears to have been the result of much highly technical investigation on the part of [the employer]. It was information which the [contractors] had neither the time nor the opportunity to obtain for themselves. It might even be doubted whether they could be expected to obtain it by their own efforts as a potential or actual tenderer. But it was indispensable information if a judgment were to be formed as to the extent of the work to be done.'

Similarly, in *Cana Construction Co Ltd* v. *The Queen* (1973), the employer was held liable for a consultant engineer's inaccurate estimate of the cost of supplying and installing mail-handling equipment given to a main contractor who relied on it in his pricing, although the exact basis of the decision is not clear.

Breach of implied warranty

Misinformation given to the contractor on the employer's behalf might also give rise to a claim for breach of an implied warranty. This is illustrated by *Bacal Construction (Midlands) Ltd* v. *Northampton Development Corporation* (1976), where the contractors submitted as part of their tender sub-structure designs and detailed priced bills of quantities for six selected blocks of dwellings in selected foundation conditions. These formed part of the contract documents, and had been prepared on the assumption that the soil conditions were as indicated on the relevant borehole data provided by the employer. The design was adequate on that assumption.

The employer's tender documents stated that the site was a mixture of Northamptonshire sand and upper lias clay. Tufa was discovered in areas of the site as work progressed, and as a result the foundations had to be redesigned and additional work had to be carried out. The Court of Appeal held that the contractors were entitled to recover compensation for breach of an implied warranty by the employer that the ground conditions would accord with the hypotheses upon which they had been instructed to design.

It is important to appreciate the difference between a misrepresentation, which involves a failure to describe the facts correctly at the time the statement is made, and a breach of a term of the contract, which relates to the time at which the contract is performed, or should be performed. The distinction was a peripheral issue in *Strachan and Henshaw Ltd* v. *Stein Industrie (UK) Ltd and Another* (1997). Strachan and Henshaw alleged that a statement had been made at a pre-contract meeting that a clocking station and tea cabins for its workforce would be situated near to the place of work. In the event it was required to place its cabins and clocking station half a mile away, with a substantial increase in walking time. It complained that this was both a breach of contract and a misrepresentation. It was held that the facts did not support a misrepresentation claim because there was nothing incorrect about the representation when it was made.

If the facts in *Strachan and Henshaw* had supported a misrepresentation claim it would have failed, because the relevant sub-contract incorporated the Model Form of General Conditions of Contract 'MF/1'. Clause 44.4 of that standard form includes a term to the effect that all the rights obligations and liabilities of the parties are fully set out in the contract. The very wide terms of the clause were held to be sufficient to exclude claims for misrepresentation.

Although some of the cases discussed provide examples of situations of potential tortious liability, it must be emphasised that since the decision of the House of Lords in *Murphy* v. *Brentwood District Council* (1990) the possibility of recovering economic (i.e. pure financial) loss in a negligence action is very restricted.

Economic loss

However, although it is now clear that normally a plaintiff cannot recover economic loss in an action for negligence, but must establish actual death or physical injury to persons or property other than the defective structure itself, economic loss can be recovered 'where there is a special relationship amounting to reliance by the plaintiff on the defendant or where the economic loss is truly consequential upon actual physical injury to person or property': *Keating on Building Contracts*, 6th edn, 1995, p170.

A 'reliance situation' will seemingly only arise if there is a special relationship of both proximity and reliance between the parties. Where the representor has some special skill or knowledge about which he gives advice, and he knows, or it is reasonably foreseeable, that the other will rely on the advice, and the advice has been acted on by the other, the resultant economic loss will be recoverable provided that it is foreseeable: *Hedley Byrne & Co Ltd* v. *Heller & Partners Ltd* (1963).

The House of Lords recently reconsidered and restricted the criteria for the special relationship. The current position appears to be that a special relationship will be considered to exist if:

- the advice is required for a purpose which is made known to the misrepresentor at the time the advice is given; and
- the misrepresentor knows or can reasonably foresee that the advice will be communicated to the other either personally or as a member of an ascertainable class in order to be used for the purpose initially made known; and
- it is known by the misrepresentor that the advice is likely to be acted upon by the other without further enquiry; and
- it is so acted upon and the other suffers some detriment: *Caparo Industries plc* v. *Dickman and others* (1990).

Because special skills and advice are features of this kind of liability, it is often associated with professional advice although, as already mentioned, today it seems that in principle the existence of a contract between the parties is in general likely to preclude recovery in negligence: *Hiron* v. *Pynford Services Ltd* (1991).

At one time the case of *Junior Books Ltd* v. *The Veitchi Co Ltd* (1982) appeared to open the floodgates to the recovery of economic loss. However in subsequent cases it has been distinguished almost to the point of extinction (e.g. *Muirhead* v. *Industrial Tank Specialities Ltd* (1986)) although it has not been directly overruled and, indeed, in *Murphy* v. *Brentwood District Council* (1990) was explained by some members of the House of Lords as resting upon the *Hedley Byrne* principle of reliance, although on what basis that 'reliance' arose is not easy to see. The House of Lords appeared to treat the employer/ nominated sub-contractor relationship as an almost unique situation.

There was no collateral contract involved in *Junior Books* between employer and nominated sub-contractor; and the relationship between a nominated sub-contractor and an employer is hardly, on any reasonable view, a 'special relationship'. It is thought that the decision is not of general application and would certainly not be extended.

The current position as determined by the House of Lords in the important case of *Murphy* v. *Brentwood District Council* (1990) may, in the context of the construction industry, be broadly summarised as follows.

- Negligence that results in a defect in the building or structure itself is not actionable in tort. There is no actionable damage to 'other property'.
- To be actionable, the defective structure must cause damage to other property or result in death or personal injury.
- If, however, the defect is discovered before it has caused damage, the cost of making good the defect is not recoverable.
- In a complex structure, it is not permitted to consider the building or structure as a series of segments, one causing damage to another – the so-called 'complex structure' theory. This was emphasised in *Tunnel Refineries Ltd* v. *Bryan Donkin Co Ltd and others* (1998).
- However, there is possibly potential liability where 'some distinct item incorporated in the structure ... positively malfunctions so as to inflict positive damage on the structure in which it is incorporated'. In the light of recent authorities such as the *Tunnel Refineries* case, this argument will only succeed where the distinct item is very distinct indeed.

6.4 Claims against the engineer

If the contractor is unable for one reason or another to recover against the employer, perhaps because of insolvency, it is not uncommon for him to consider making a claim against the engineer. Although the engineer is named in the contract, and is required to carry out his functions under the contract, he is not a party to the contract, and the contractor has no contractual relationship with him.

The leading case on this question is *Pacific Associates Inc* v. *Baxter* (1988), where contractors were unsuccessful in an action for negligence against a supervising engineer under a FIDIC contract who had failed to certify certain sums allegedly due to them. The contractors also alleged unsuccessfully a breach by the engineer of a duty to act impartially.

The Court of Appeal, in dismissing the contractors' claim, stressed the importance of the terms of the contract between the employer and the contractor which provided for arbitration of disputes arising under the contract. The logic of the decision was as follows.

- The engineer was an agent of the employer.
- The engineer had a contractual obligation to the employer to use skill and care and to act fairly between the parties.
- The contractors had relied on its remedies against the employer under the contract between them by going to arbitration over the disputed claims. In the event, the arbitration was settled on terms of an *ex gratia* payment by the employer to the contractor who was attempting to recover the shortfall between the amount claimed and the amount of the settlement by means of an action in tort against the engineer.
- The engineer had not assumed responsibility to the contractor for economic loss resulting from a breach of any of the obligations in the contract between the employer and the contractors.
- Therefore, there was no basis, either on the *Hedley Byrne* principle or otherwise, by which the engineer could be said to owe a duty of care to the contractor.
- There was an arbitration provision that enabled the contractor to recover from the employer. The engineer had a duty to act in accordance with the construction contract, but that duty arose from the contract between the engineer and the employer. The contractor could challenge the performance of the engineer by claiming against the employer. There was, therefore, no justification for imposing on the contractual structure an additional liability in tort.

There was an additional problem for the contractor in *Pacific Associates* in that there was a term of the contract that the engineer would not be liable to the contractor in respect of his actions, but most commentators on the case conclude that the result would have been the same even without that term. (See also section 2.2.)

6.5 Failure to issue certificates

Both the ICE Conditions and the Minor Works Conditions require the engineer to issue certificates of payment to the contractor, the duty being spelled out in clause 60(2) of the Sixth Edition and clause 7.3 of the Minor Works Conditions. The engineer's failure to do so is a breach of contract for which the employer may be liable. The engineer's certificate is in both cases a condition precedent to payment under the contract, but it is clear that the courts will protect the contractor's rights even in the absence of a certificate: *Croudace Ltd* v. *London Borough of Lambeth* (1986).

There are many instances where a contractor has been able to recover without a certificate, sometimes on the basis of an implied term (as in *Panemena Europea Navigacion Lda* v. *Frederick Leyland & Co Ltd* (1947)).

If the engineer under-certifies, or the certificate contains an error, it is thought that the contractor's remedy is to invoke the disputes procedure in clause 66 of the Sixth Edition, and the employer meets his contractual obligation by paying the amount certified by the engineer, even if this is erroneous: *Lubenham Fidelities & Investment Co* v. *South Pembrokeshire District Council and Wigley Fox Partnership* (1986). This was yet another case arising under the JCT Standard Form of Building Contract, 1963 edition, where the architect made wrongful deductions, and the employer refused to pay sums in excess of those certified, the Court of Appeal holding that it was not in breach of contract by so doing.

Both clause 60(6) of the ICE Conditions and clause 7.3 of the Minor Works Conditions give the contractor a right to interest in the event of failure by the engineer to certify and so his rights are adequately safeguarded.

CHAPTER SEVEN
PROCEDURES AND EVIDENCE

7.1 Introduction

The contract gives the contractor the right to recover cost in certain circumstances and, of course, the contractor may also complain of breaches of contract by the employer. The word 'claim' might be used to describe either. In this chapter we will concentrate on the former.

If the contractor wishes to make a claim under the contract he must operate the machinery that it provides. As we have seen, that machinery is not exhaustive of the contractor's remedies but by relying on the contractual mechanism the contractor hopes to get the benefit of early reimbursement of the 'cost' he has incurred as a result of the specific event provided for under specific clauses of the contract.

A revised and extended definition of 'cost' is provided in clause 1(5) of the Sixth Edition:

> The word 'cost' when used in the Conditions of Contract means all expenditure properly incurred or to be incurred whether on or off the Site including overhead finance and other charges properly allocatable thereto but does not include any allowance for profit.

The addition of the words 'to be incurred' overcomes the difficulty which arose under the Fifth Edition where there was no reference to costs 'to be incurred' as there is in the Sixth Edition. Under the Sixth Edition it is not necessary to wait until all costs have been spent before making the claim, whereas arguably this was a requirement under the Fifth Edition.

The word 'properly' in the definition is intended to limit what is recoverable to those heads of claim allowable under the rules in *Hadley* v. *Baxendale* (1854) so that common law principles of 'remoteness' are applicable. It also implies that only costs that are, in whole or in part, incurred *necessarily* are claimable. (See section 8.1.)

The clause 2(5) definition in the Fifth Edition was more terse and enigmatic. It defined 'cost' as including 'overhead costs whether on or off the Site except where the contrary is expressly stated'. It should be noted in particular that finance charges were not expressly included in the Fifth Edition definition.

The ICE Conditions do not make any special provision for the recovery of prolongation or disruption costs. The cost of prolongation or disruption is recoverable as part of 'cost' if it can be shown to result from one of the specific events that leads to payment. This is often very difficult to establish in practice. It is not sufficient merely to establish that there has been delay as a result of one of the specified events, and that there have been costs arising out of delay. The cost must be shown to have resulted from the specified event. It is often the case that a number of factors have contributed to the delay, some of which entitle the contractor to recover and some do not. If they affected progress concurrently it is necessary to establish which factor was the dominant cause. This is a question of fact which can be extremely difficult to establish. Common sense sometimes provides the answer, but sophisticated computer analysis of anticipated and actual programme and resource levels may also be required.

Clause 13(3) of the Fifth Edition was particularly important because it was a general provision for the recovery of cost 'beyond that reasonably to have been foreseen by an experienced contractor at the time of tender' incurred as a result of engineer's instructions or directions. Neither of these terms was defined, and it seemed that there was no contractual restriction as to what amounts to an instruction or direction for the purposes of clause 13(3). That subclause was, therefore, a fertile source of claims.

In the Sixth Edition clause 13(3) has been amended significantly. It now provides that if, as a result of engineer's instructions which involve the contractor in delay or disrupt his arrangements or methods of construction so as to incur cost beyond that reasonably to have been foreseen by an experienced contractor at the time of tender then the engineer shall take such delay into account in determining any extension of time under Clause 44 and, subject to compliance with the notice requirements of clause 52(4) the contractor is entitled to recover 'the amount of such cost as may be reasonable *except to the extent that such delay and extra cost result from the Contractor's default'*. Profit is to be added to cost only in respect of any additional permanent or temporary work.

7.2   *The ICE Conditions – notice of claims*

Clause 52(4) provides as follows:

(a)   If the Contractor intends to claim a higher rate or price than one notified to him by the Engineer pursuant to sub-clauses (1) and (2) of this Clause or Clause 56(2) the Contractor shall within 28 days after such notification give notice in writing of his intention to the Engineer.

(b)   If the Contractor intends to claim any additional payment pursuant to any Clause of these Conditions other than sub-clauses (1) and (2) of this Clause or Clause 56(2) he shall give notice in writing of his intention to the Engineer as soon as may be reasonable and in any event within 28 days after the happening of the events giving rise to the claim. Upon the happening of such events the Contractor shall keep such contemporary records as may reasonably be necessary to support any claim he may subsequently wish to make.

(c)   Without necessarily admitting the Employer's liability the Engineer may upon receipt of a notice under this Clause instruct the Contractor to keep such contemporary records or further contemporary records as the case may be as are reasonable and may be material to the claim of which notice has been given and the Contractor shall keep such records. The Contractor shall permit the Engineer to inspect all records kept pursuant to this Clause and shall supply him with copies thereof as and when the Engineer shall so instruct.

(d)   After the giving of a notice to the Engineer under this Clause the Contractor shall as soon as is reasonable in all the circumstances send to the Engineer a first interim account giving full and detailed particulars of the amount claimed to that date and of the grounds upon which the claim is based. Thereafter at such intervals as the Engineer may reasonably require the Contractor shall send to the Engineer further up to date accounts giving the accumulated total of the claim and any further grounds upon which it is based.

(e)   If the Contractor fails to comply with any of the provisions of this Clause in respect of any claim which he shall seek to make then the Contractor shall be entitled to payment in respect thereof only to the extent that the Engineer has not been prevented from or substantially prejudiced by such failure in investigating the said claim.

(f)   The Contractor shall be entitled to have included in any interim payment certified by the Engineer pursuant to Clause 60 such amount in respect of any claim as the Engineer may consider due to the Contractor provided that the Contractor shall have supplied sufficient particulars to enable the Engineer to determine the amount due. If such particulars are insufficient to substantiate the whole of the claim the Contractor shall be entitled to payment in respect of such part of the claim as the particulars may substantiate to the satisfaction of the Engineer.

Clause 52(4) is a general notice provision that lays down the procedure for claims by the contractor for extra payment arising under any clause in the contract and not merely under clause 52 itself. All the contract clauses that give the contractor a right to claim additional money are subject to clause 52(4). It is a comprehensive requirement for notice and the provision of information by the contractor. However, as will be seen, the contractor's failure to comply with its provisions merely bars his claim 'to the extent that the Engineer has ... been prevented from or substantially prejudiced by [the contractor's non-compliance] in investigating the ... claim'. Moreover clause 52(4) applies only to claims made under the contract. It does not apply to common law claims, for example those for breach of contract. The contractor is not contractually bound to give notice, but he must do so if he wishes to claim under the Conditions and it is a condition precedent to his right to claim reimbursement in interim certificates.

7.2.1   Giving notice

Clause 52(4)(a) covers claims for higher rates and prices other than those notified to the contractor by the engineer under clauses 52(1) or 52(2) (ordered variations) or clause 56(2) (increase or decrease of rates). In such a case, the contractor must give written notice to the engineer of his intention to claim a higher rate or price within 28 days 'after ... notification' by the engineer of the rate or price. The important date is the date of actual receipt by the contractor of the notification.

The notice of intention to claim in general terms is valid so long as it specifies that a claim is being made and identifies in general terms the varied work, etc. to which it relates: *Tersons Ltd* v. *Stevenage Development Corporation* (1963).

Whilst the giving of notice is expressed to be a condition precedent to the right to claim, clause 52(4)(e) applies, discussed below.

Clause 52(4)(b) covers all claims for additional payment arising under any other clause of the conditions. The contractor must give written notice to the engineer (under the Sixth Edition) 'as soon as may be reasonable *and in any event within 28 days after the happening of the events giving rise to the claim*'. Notice after a long lapse of time is disputable even under the Fifth Edition but the italicised phrase did not appear in the that version. It is intended to bar late claims, and it is essential that contractors' site staff appreciate the importance of this requirement.

In *Tersons Ltd* v. *Stevenage Development Corporation* (1963) the Court of Appeal gave a generous interpretation to the words 'at the earliest possible opportunity', and Lord Justice Upjohn emphasised that where notice is required to be given 'as soon as is reasonable' this may give more latitude than where it is required to be given 'as soon as is practicable'. Under clause 52(4)(b) of both editions, the question of reasonableness arises; whether or not a notice is given at the right time is a question of fact, and a matter for the discretion of the engineer or arbitrator. It should be noted that the 'as soon as may be reasonable' requirement and the 'within 28 days' stipulation are not alternatives. Notice might be given within 28 days and yet be late because it has not be given 'as soon as may be reasonable'.

Contractors should bear in mind that the purpose of the notice requirement is not merely to trigger the contractual machinery for reimbursement of costs. It is also to enable the engineer to investigate the claim and, if necessary, take remedial action.

The decision of Mr Justice Donaldson in *J Crosby & Sons Ltd* v. *Portland Urban District Council* (1967) is of some assistance in properly interpreting clause 52(4). It emphasises not only that the notice provisions are a condition precedent to the contractor's right to recover under the provisions of the contract, but that one of the aims of the provisions is to provide the engineer with the opportunity of investigating the claim when or soon after the events giving rise to it occur, and to advise the employer of the cost implications and of possible remedial action. Mr Justice Donaldson, in considering clause 40(1) of the ICE Conditions, Fourth Edition, referred to the lack of notice of intention to make a claim consequent on a suspension of work. He said:

> 'The object of the proviso is to bring the engineer's attention to the fact that the continuation of the suspension may be costly to his employer. This object is worthy of achievement and can be achieved by the giving of a general notice whatever the actual or anticipated circumstances of the suspension.'

As the contractor had given no notice within the period of 28 days required by clause 40(1) his claim failed. Exactly the same principles apply to clause 52(4)(a) and (b).

7.2.2 Records

The second limb of clause 52(4)(b) is also important because it requires the contractor 'upon the happening of [the events giving rise to the claim]' to keep 'such contemporary records as may reasonably be necessary to support any claim he may subsequently wish to make'. In other words, the contractor is bound to keep contemporary records immediately the relevant claims events occur, even if this is before the time when it is reasonable to give notice of intention to claim. The sub-clause is silent about the sort of records required to be kept, except that they must be 'reasonably . . . necessary to support' any subsequent claim. They should cover labour, plant and all other costs. The provisions of clause 52(3) dealing with daywork give some indication of the level of detail required.

Clause 52(4)(c) also deals with records. The engineer may require the contractor to keep such contemporary records as are reasonable and may be material to the claim that has been notified. The engineer is to be permitted to inspect those records, and can require the contractor to provide copies.

There is no express contractual sanction if the contractor fails to meet these requirements, though his failure to do so is a breach of contract. Only an unwise contractor will neglect to observe the provision since this will, in practice, limit the amount of his claim. Figures cannot be plucked out of the air and good contemporaneous records are essential.

It is suggested that the engineer may require the contractor to keep records in a particular form under clause 52(4)(c) provided, of course, he acts reasonably. Computerised records kept on a particular software system would probably be a reasonable requirement on larger projects and the contractor may well be required to keep records in such a form as would tie in cost with the event relied on so as to establish a chain of cause and effect. Agreement between contractor and engineer as to the sort and form of records required can be very time-saving and avoid a common source of disputes. The records must be kept at the contractor's own expense.

The prudent contractor does not need to be told, either by the terms of the contract or by the engineer, to keep detailed records. Without greatly adding to his administrative cost a full computer database can be maintained which will provide necessary information to calculate entitlement to additional payment for variations and at the same time enable a proper presentation to be made in support of claims for extensions of time and extra cost. The expense of preparing a fully supported claim is then minimal, and the chances of success very greatly improved.

7.2.3 Claims procedure

The actual claims procedure is set out in clause 52(4)(d). This requires the contractor, after giving notice to the engineer, to send to the engineer 'a first interim account giving full and detailed particulars of the amount claimed' up to the date of the interim account and setting out the grounds on which the claim is based. Although it is the engineer's duty to determine the amount due, the contractor as the person in possession of the facts and figures must provide them. It is helpful if the claim is backed by relevant supporting evidence.

The interim account is to be submitted 'as soon as is reasonable in all the circumstances' after the notice has been given to the engineer. The interim account must refer to costs actually incurred and may also refer to prospective costs. The contractor submits further up-to-date accounts at intervals reasonably required by the engineer amending anticipated future costs as they are incurred and become actual costs. In most cases these accounts will tie in with the timing of certifications.

It is to be noted that the subsequent accounts must give the 'accumulated total' of the contractor's claim and any further grounds on which it is based. As we have seen, clause 52(4)(d) requires the contractor to set out the grounds on which his claim is based when he sends his first interim account, and so the reference to 'further grounds' must be to grounds that subsequently come to light.

It is in the contractor's interest to state the figures accurately and not to make an excessive claim in the mistaken belief that the engineer will then pay the greater part of it without verifying the figures. An inaccurate account may well prejudice the engineer in investigating the claim (see clause 52(4)(f)). The engineer may, and indeed must, come back to the contractor for back-up information if necessary. The contemporaneous records already referred to are therefore vitally important.

There appears to be no time limit on the engineer's investigation, but it is suggested that the engineer is under an implied duty to investigate the contractor's claim within a reasonable time of its submission: *Croudace Ltd* v. *London Borough of Lambeth* (1986).

Obviously with the issue of the final certificate the engineer becomes *functus officio,* but if the contractor is able to provide further evidence in support of his claim he may re-submit the claim at any time before the final certificate. The matter is also reviewable under clause 66. In practice, if the disputes procedure is invoked the engineer considers the matter again. In that case both the engineer and adjudicator or arbitrator are bound to consider all information that was made available to the engineer before the final certificate.

7.2.4 Failure to comply with contractual provisions

Clause 52(4)(e) deals with the contractor's failure to comply with the notice and procedural provisions. His failure to do so means that he is entitled to payment 'only to the extent that the Engineer has not been prevented from or substantially prejudiced by such failure in investigating the claim'. This limitation covers the whole of the clause 52(4).

It is a question of fact whether the engineer 'has ... been prevented ... or substantially prejudiced'. It is easy to see, though, that late notification of delaying events or alleged cost may well make it more difficult for the engineer to investigate the claim, and it is for the contractor to demonstrate that the engineer has not been prevented or prejudiced. It is therefore extremely dangerous to rely on clause 52(4)(e) to escape the consequences of failure to comply with the procedural requirements of clause 52.

7.2.5 Payment

The contractor's right to payment of verified claims is set out in clause 52(4)(f). Any amounts determined by the engineer to be due to the contractor are to be included in interim certificates, provided that the contractor has provided sufficient particulars. The engineer's duty is to determine any amount due to the contractor. If the particulars that the contractor has provided are insufficient to substantiate the whole of the claim, he is entitled to payment of such part of it as he has substantiated to the engineer's satisfaction.

If the contractor substantiates his claim, then the engineer must certify it – it is a contractual right. The engineer's decision is, of course, reviewable in arbitration and adjudication.

Clause 52(4) is essentially practical in its operation, which depends on having established and clear procedures for the keeping of accurate records that are capable of pinpointing the actual costs and what is happening on site. The counsel of perfection is that contemporaneous records should be agreed.

7.3 The Minor Works Conditions – clause 6

Clause 6 of the Minor Works Conditions reads:

> 6.1 If the Contractor carries out additional works or incurs additional cost including any cost arising from delay or disruption to the progress of the Works as a result of any of the matters referred to in paragraphs (a) (b) (d) (e) or (f) of Clause 4.4 the Engineer shall certify and the Employer shall pay to the Contractor such additional sum as the Engineer after consultation with the Contractor considers fair and reasonable. Likewise the Engineer shall determine a fair and reasonable deduction to be made in respect of any omission of work.
>
> 6.2 In determining a fair and reasonable sum under Clause 6.1 for additional work the Engineer shall have regard to the prices contained in the Contract.

Clause 6.1 does not require the contractor to give notice of claim under its provisions as such, but clause 4.4 to which it refers does provide for the contractor to make a 'written request' to the engineer for an extension of time. Furthermore, clause 7.2 requires the contractor to submit to the engineer monthly statements showing the value of the work executed, the value of any goods and materials delivered to the site 'and any other items which the contractor considers should be included in an interim certificate'. This is apt to cover additional payments under clause 6.1 as well as such things as the value of temporary works or constructional plant on the site.

'Cost' is defined in condition 1.3 in terms similar to the equivalent in the Sixth Edition.

Clause 6.1 covers not only additional works but also additional cost incurred by the contractor including, but not limited to, prolongation or disruption costs incurred as a result of:

- Variation instructions (clause 2.3(a)).
- Instructions requiring suspension of the whole or part of the works (clause 2.3(c)).

- Instructions changing the intended sequence of the works (clause 2.3(d)).
- Instructions requiring testing and investigation where the test or investigation does not disclose non-compliance with the contract (clause 2.3(b)).
- Delayed receipt by the contractor of necessary instructions, drawings or other information (clause 4.4(d)).
- Employer's failure to give adequate access to the works or possession of land required for their performance (clause 4.4(e)).
- Delayed receipt by the contractor of materials to be provided by the employer under the contract (clause 4.4(f)).

Although not expressly stated, it is clear that the contractor must substantiate the amount of any additional payment to which he claims to be entitled, and that these sums, when determined, are to be included by the engineer in interim certificates.

Clause 6.1 very sensibly recognises the realities of a claims situation, because the engineer is to certify additional payments 'after consultation with the Contractor' and the only practical issue is what sum the engineer 'considers fair and reasonable'. The contractor must clearly co-operate with the engineer in giving particulars of the cost involved, and must respond to any requests for information from the engineer.

If the contractor is unhappy with the sum certified in respect of his claim he can refer the matter to adjudication or conciliation (see sections 7.5 and 7.6 below).

7.4 Evidence

Although the contracts provide for the making of financial claims by the contractor and for their settlement under the contract provisions, it is for the contractor to substantiate his claim by means of proper evidence so as to demonstrate and prove to the satisfaction of the engineer his entitlement to the amount claimed. 'He who asserts must prove' is a basic legal maxim and the standard of proof is 'the balance of probabilities'. As we have seen, clause 52(4)(d) of the ICE Conditions requires the contractor to provide the engineer with 'full and detailed particulars of the amount claimed ... and of the grounds upon which the claim is based'. The contractor is, therefore, required to provide documentary and other evidence justifying the amount claimed as well as describing the basis of his claim.

The primary evidence will be the contractor's own contemporaneous records and, in the case of claims under the ICE
Conditions, contemporaneous records will or should be available,
both as to cost and otherwise. It is in the interest of contractors to
keep detailed records of all costs related to the individual contract
from the outset, and in a prolongation or disruption situation other
information will include comparative programmes, records of site
meetings, relevant correspondence and so on. The engineer cannot
be, and is not, expected to determine the amount of the claim in a
vacuum. A full database of all relevant information maintained
from the start of the project is invaluable.

If the contractor cannot produce sufficient particulars to substantiate the whole of his claim, he will be entitled to payment only in
respect of that part of the claim which he has substantiated to the
satisfaction of the engineer. It is true that whether or not the engineer
has been prejudiced or prevented by the lack of back-up evidence is a
matter which is arbitrable, and very considerable practical problems
arise where the contractor is seeking to recover the cost of prolongation or disruption. However, it is clear that it is for the contractor to establish the rates that are to be applied or varied.

Every claim is different on its own facts, but each and every claim
must be properly formulated and backed up. Many contractors'
claims are deficient in the matter of supporting evidence. The
contracts require that the contractor should be able to tie in the
additional cost with the event on which he relies.

This is particularly difficult in relation to overhead costs, whether
on or off site. The contractor may claim, for example, that a
reduction in the volume of work under the contract has led to a
shortfall in the recovery of overhead costs, and will often adopt the
simplistic approach that the overhead costs, which are spread over
the rates for measured work, are proportional to the time taken by
construction. In fact, this is an incorrect assumption, since on-site
overhead costs are divisible into the initial and final costs such as
erection and dismantling of plant and temporary buildings, and
those that are spread over the contract period such as plant hire.
Overhead costs are discussed in detail in Chapter 3.

The rate at which plant and equipment should be charged is the
actual cost to the contractor. If the plant is hired in from a plant
hirer, the amount is the hire rate actually paid, and even this may
raise difficulties where plant is idle during a period of prolongation.
In such circumstances, the contractor must make realistic efforts to
redeploy the plant elsewhere, or else take it off hire.

Where the contractor is using his own plant, the true cost to him must be calculated. It is not sufficient to take the rates at which the plant would have had to be hired in had it not been the contractor's own property, and effectively his claim will be limited to the cost of depreciation, the cost of capital and any necessary additional maintenance. This was the logical approach of the Court of Appeal in 1940 (*Bernard Sunley & Co Ltd* v. *Cunard White Star Ltd*) and also of the Official Referee in a building case in 1995 (*Alfred McAlpine Homes North Ltd* v. *Property and Land Contractors Ltd*). Hire charges might be justified, depending on the facts, if the contractor is able to show that as a result of delays on the contract in question the plant had not been released for use elsewhere, and that consequently he had had to hire plant for the other contract.

The sort of evidence required by the engineer will depend on the nature of the claim. Some practical examples are given in section 8.3 below.

7.5 Disputes procedures – conciliation and adjudication

ICE Contracts all contain comprehensive disputes clauses defining when a dispute is considered to have arisen and how it should be dealt with. Until May 1998 it was generally considered that the only way in which an engineer's decision could be overturned was through the machinery of clause 66. If settlement could not be achieved by negotiation, either informal or structured (for example by using the ICE conciliation procedure) a challenge to the engineer's decision had to be taken to formal arbitration. The courts had no part to play and were unable to interfere with the engineer's decisions. This was all changed by two events which occurred within a period of three weeks.

On 1 May 1998 Part II of the Housing Grants, Construction and Regeneration Act 1996 came into force. Parties to construction contracts formed on or after that date were given the right to refer disputes to a remarkable new procedure called adjudication. If the contract did not expressly give that right, the statute filled the gap. The ICE introduced amendments to its contracts to cover the point, although there is real doubt about whether those amendments satisfy the requirements of the Act. The right is there, though, either under the contract or under the Act.

Shortly after the Act came into force the House of Lords delivered judgment in *Beaufort Developments (NI) Ltd* v. *Gilbert-Ash (NI) Ltd*. It was held that the courts had power to open up and review certificates given by the architect in a building contract, and the logic of the decision makes it clear that the courts have a similar power to open up the engineer's decision in a civil engineering contract. This does not mean that a contractor can take a unilateral decision to proceed with a claim in court rather than arbitration. If there is an agreement to arbitrate, as in the ICE contracts, either party will be able to have court proceedings stayed under section 9 of the Arbitration Act 1996 but he will have to make prompt application to the court to do so. In some cases the defendant in the court action may elect not to apply for a stay either because he would prefer the court to deal with the dispute or because he wishes to involve third parties with whom he does not have an arbitration agreement.

Arbitration and court proceedings are outside the scope of this work, but conciliation is very much a part of the contractual provisions relating to civil engineering claims, and adjudication may well become an integral part of the claims process.

7.6 *Conciliation and adjudication under the ICE Conditions*

The relevant parts of clause 66 provide as follows:

(1) In order to overcome where possible the causes of disputes and in those cases where disputes are likely still to arise to facilitate their clear definition and early resolution (whether by agreement or otherwise) the following procedure shall apply for the avoidance and settlement of disputes.

(2) If at any time
(a) the Contractor is dissatisfied with any act or instruction of the Engineer's Representative or any other person responsible to the Engineer or
(b) the Employer or the Contractor is dissatisfied with any decision opinion instruction direction certificate or in connection with the Contract or the carrying out of the Works
the matter of dissatisfaction shall be referred to the Engineer who shall notify his written decision to the Employer and the Contractor within one month of the reference to him.

(3) The Employer and the Contractor agree that no matter shall constitute nor be said to give rise to a dispute unless and until in respect of that matter
(a) the time for the giving of a decision by the Engineer on a matter of dissatisfaction under Clause 66(2) has expired or the decision given is unacceptable or has not been implemented and in

consequence the Employer or the Contractor has served on the other and on the Engineer a notice in writing (hereinafter called the Notice of Dispute) or

(b)   an adjudicator has given a decision on a dispute under Clause 66(6) and the Employer or the Contractor is not giving effect to the decision, and in consequence the other has served on him and the Employer a Notice of Dispute

and the dispute shall be that stated in the Notice of Dispute. For the purposes of all matters arising under or in connection with the Contract or the carrying out of the Works the word 'dispute' shall be construed accordingly and shall include any difference.

(4)   (a)   Notwithstanding the existence of a dispute following the service of a Notice under Clause 66(3) and unless the Contract has already been determined or abandoned the Employer and the Contractor shall continue to perform their obligations.

(b)   The Employer and the Contractor shall give effect forthwith to every decision of

(i)   the Engineer on a matter of dissatisfaction given under Clause 66(2) and

(ii)   the adjudicator on a dispute given under Clause 66(6)

unless and until that decision is revised by agreement of the Employer and Contractor or pursuant to Clause 66.

(5)   (a)   The Employer or the Contractor may at any time before service of a Notice to Refer to arbitration under Clause 66(9) by notice in writing seek the agreement of the other for the dispute to be considered under the Institution of Civil Engineers' Conciliation Procedure (1994) or any amendment or modification thereof being in force at the date of such notice.

(b)   If the other party agrees to this procedure any recommendation of the conciliator shall be deemed to have been accepted as finally determining the dispute by agreement so that the matter is no longer in dispute unless a Notice of Adjudication under Clause 66(6) or a Notice to Refer to arbitration under Clause 66(9) has been served in respect of that dispute not later than 1 month after receipt of the recommendation by the dissenting party.

(6)   (a)   The Employer and the Contractor each has the right to refer a dispute as to a matter under the Contract for adjudication and either party may give notice in writing (hereinafter called the Notice of Adjudication) to the other side at any time of his intention to do so. The adjudication shall be conducted under the Institution of Civil Engineers' Adjudication Procedure (1997) or any amendment or modification thereof being in force at the time of the said Notice.

(b)   Unless the adjudicator has already been appointed he is to be appointed by a timetable with the object of securing his appointment and referral of the dispute to him within 7 days of such notice.

(c) The adjudicator shall reach a decision within 28 days of referral or such longer period as is agreed by the parties after the dispute has been referred.

(d) The adjudicator may extend the period of 28 days by up to 14 days with the consent of the party by whom the dispute was referred.

(e) The adjudicator shall act impartially.

(f) The adjudicator may take the initiative in ascertaining the facts and the law.

(7) The decision of the adjudicator shall be binding until the dispute is finally decided by legal proceedings or arbitration (if the contract provides for arbitration or the parties otherwise agree to arbitration) or by agreement.

(8) The adjudicator is not liable for anything done or omitted in the discharge or purported discharge of his functions as adjudicator unless the act or omission is in bad faith and any employee or agent of the adjudicator is similarly not liable.

7.6.1 Conciliation

Conciliation is quite different from adjudication. It is a process which is unaffected by statute, and it depends upon the voluntary participation of the parties. It is not aimed at establishing the entitlements of the parties, even on an interim basis. Its object is to facilitate the settlement of the dispute upon terms that are acceptable to both parties.

Before the contractual process of conciliation can begin, there must be a dispute as defined by the contract. Clause 66(3) states that there is no dispute unless and until a matter of dissatisfaction has been referred to the Engineer and either he has failed to give a decision on it within one month, or the decision is unacceptable or has not been implemented. Alternatively an adjudicator's decision (see below) may have been given but ignored. In those circumstances the dissatisfied party can serve a Notice of Dispute stating the matter(s) in dispute.

The service of a formal Notice of Dispute may seem to be an inconvenient additional bureaucratic requirement, but its intention is to clarify the matters in dispute. Clarification of the matters in dispute is an essential step towards their resolution.

The formal Notice of Dispute having been given, either party can suggest that the dispute be referred to conciliation using the ICE procedure. Neither party can insist on conciliation. The parties may feel that the process will give an opportunity to demonstrate the common sense merits of the case quickly and cheaply, avoiding the cost and delay of arbitration.

If the process is to be used in the resolution of a claim dispute, the contractor will set out in a Notice of Conciliation the matters in dispute and remedy sought. That notice is served on the employer. The conciliator may have been agreed in advance, but if not the contractor will probably give three or four names of people who might be appointed to act in that capacity. If the parties cannot agree, an application can be made to the President of the ICE to appoint a conciliator.

The conciliator will wish to deal with the conciliation procedures within a relatively short period – two months at the most. He will adopt any procedure that he considers appropriate leading to a meeting or series of meetings attended by senior representatives of the parties with authority to make final decisions. Each party might attend such meetings with several representatives – often with those closely involved in the subject of the dispute, senior management and legal or technical advisers.

At some stage in the process, possibly before the first meeting, the conciliator will require the responding party to set out briefly the nature of its case on the matter in dispute. He may call for production of documents and may visit the site if he thinks it will be helpful.

Typically the meetings with the parties will involve full sessions, with both parties present, and then private meetings with one party at a time. The conciliator will investigate not just the merits of the case being put to him, but also possible commercial approaches to settlement. The parties will be encouraged to discuss in confidence their opinions of the case and their overall objectives.

The conciliator will encourage the making of settlement proposals and will assist the process of negotiation, but will not make a decision about the rights of the parties. If a settlement is not achieved he will make a recommendation, which may or may not form the basis of a settlement. If his recommendation is not accepted, and the parties take their dispute to arbitration or litigation, the matters discussed in the conciliation meetings will remain confidential and the conciliator will not be called by either party as a witness. He will not be able to accept appointment as arbitrator in connection with the same dispute.

If on the other hand settlement is achieved the terms will be incorporated in a formal agreement, bringing the dispute to a close.

It is likely that very substantial costs will have been saved by both parties.

7.6.2 Adjudication

'Adjudication' is not a new word. It was a feature of the New Engineering Contract, or Engineering and Construction Contract as it came to be known, from the start. It was also an important safeguard for sub-contractors working under the various standard forms of sub-contract used with JCT contracts, giving them protection against unjustified deductions from interim payments. Adjudication had also been introduced into the JCT form of Design and Build Contract, although it is seldom used.

Housing Grants, Construction and Regeneration Act 1996

None of these procedures has much in common with the remarkable adjudication process required to be incorporated in all construction contracts by Part II of the Housing Grants, Construction and Regeneration Act 1996.

The relevant provisions of the Act came into force on 1 May 1998. They apply to construction contracts, which are defined by section 104(1) of the Act as agreements

'with a person for any of the following:

(a) the carrying out of construction operations;
(b) arranging for the carrying out of construction operations by others, whether under sub-contract to him or otherwise;
(c) providing his own labour, or the labour of others, for the carrying out of construction operations.'

The definition also includes agreements to carry out architectural, design or surveying work, or to provide advice on building, engineering, interior or exterior decoration or on the laying-out of landscape in connection with construction operations.

The term 'construction operations' is defined at length. Simply put, it covers any activity which is normally the subject of a civil engineering contract or sub-contract.

Drilling for, or extraction of, oil and natural gas are excluded, as is mineral extraction and ancillary works. The assembly, installation

185

and demolition of plant and machinery, and erection or demolition of steelwork for the purposes of supporting or providing access to plant and machinery are excluded, but only on a site where the primary activity is nuclear processing, power generation or water or effluent treatment, or the production, transmission, processing or bulk storage of chemicals, pharmaceuticals, oil, gas, steel or food and drink.

There is also an exception in the case of a contract with a residential occupier, and the Act only applies to a contract in writing. The definition of 'in writing' is, however, extremely wide. If an agreement is 'evidenced in writing', although not formally drawn up and signed, it is treated as being in writing.

The Act's provisions with regard dispute resolution are set out in section 108:

'(1)   A party to a construction contract has the right to refer a dispute arising under the contract for adjudication under a procedure complying with this section.

For this purpose "dispute" includes any difference.

(2)   The contract shall —

    (a)   enable a party to give notice at any time of his intention to refer a dispute to adjudication;

    (b)   provide a timetable with the object of securing the appointment of the adjudicator and referral of the dispute to him within 7 days of such notice;

    (c)   require the adjudicator to reach a decision within 28 days of referral or such longer period as is agreed by the parties after the dispute has been referred;

    (d)   allow the adjudicator to extend the period of 28 days by up to 14 days, with the consent of the party by whom the dispute has been referred;

    (e)   impose a duty on the adjudicator to act impartially; and

    (f)   enable the adjudicator to take the initiative in ascertaining the facts and the law.

(3)   The contract shall provide that the decision of the adjudicator is binding until the dispute is finally determined by legal proceedings, by arbitration (if the contract provides for arbitration or the parties otherwise agree to arbitration) or by agreement.

The parties may agree to accept the decision of the adjudicator as finally determining the dispute.

(4)   The contract shall also provide that the adjudicator is not

liable for anything done or omitted in the discharge or purported discharge of his functions as adjudicator unless the act or omission is in bad faith, and that any employee or agent of the adjudicator is similarly protected from liability.

(5) If the contract does not comply with the requirements of subsection (1) to (4), the adjudication provisions of the Scheme for Construction Contracts apply.

(6) For England and Wales, the Scheme may apply the provisions of the Arbitration Act 1996 with such adaptations and modifications as appear to the Minister making the scheme to be appropriate.

For Scotland, the Scheme may include provision conferring powers on courts in relation to adjudication and provision relating to the enforcement of the adjudicator's decision.'

ICE amendments and the Scheme

Clause 66 of the ICE Conditions (Fifth and Sixth Editions) has been amended with the intention of complying with section 108 of the Act. The draftsmen have, however, tried to protect the engineer's position by providing in clause 66(3) that no matter shall be considered to be a dispute unless and until it has been referred to the engineer for a decision. It is intended that the engineer will have an opportunity to deal with the matter before any approach is made to an adjudicator.

Whilst the draftsmen's motives may have been quite reasonable, there is a real doubt as to whether they were successful. The Act requires the contract to enable a party to give notice 'at any time' of his intention to refer a dispute to adjudication (section 108 (2) (a)). 'Dispute' includes any difference (section 108 (1)). Either party should therefore have the right to submit 'any difference' to adjudication at any time, whether or not it has yet been referred to the Engineer for a decision.

If the contract does not comply with the requirements of the Act, and arguably the ICE Conditions do not, the adjudication provisions of the Scheme for Construction Contracts apply (section 108 (5)). Those provisions will certainly apply if the parties have used a copy of the contract published without the March 1998 amendments.

The Scheme for Construction Contracts was set out in a statutory instrument which also came into force on 1 May 1998. It provides a set of rules and procedures to be adopted by the parties and the adjudicator. It is comprehensive and detailed. It is clear and for the most part easy to follow. There are however very real differences between the provision of the Scheme and the ICE Adjudication Procedure (1997) which clause 66(6) says should apply. For example, the adjudicator acting under the Scheme for Construction Contracts rules cannot withhold his decision pending payment of his fees but under the ICE Adjudication Procedure he can providing that he has given the parties notice of his intention to do so at least seven days before his decision is due. Under the Scheme, either party can require the adjudicator to give reasons for his decision, whereas under the ICE procedure the adjudicator is not required to give reasons.

These differences may lead to very serious difficulty. An adjudicator appointed to act in an ICE contract must establish immediately on appointment whether the adjudication is to be conducted under the provisions of the Scheme or the ICE Adjudication Procedure. It is submitted that if there is any doubt or disagreement on the point the adjudicator should assume that the Scheme applies and act accordingly.

Enforcement

Once given, the adjudicator's decision is binding on an interim basis (section 108(3)). A sum that the adjudicator has found is due to the contractor, whether as part of the contract value or as a valid claim, becomes payable as a contractual entitlement. Failure to pay this, or any other sum due under the contract, may give rise to a right to suspend work under section 112 of the Act.

Enforcement may potentially be effected in one of two ways. Clause 66(9)(a) excludes a dispute about a failure to comply with an adjudicator's decision from the ambit of the arbitration agreement.

The contractor can issue court proceedings claiming the sum that the adjudicator has decided is due. Typically this will be followed by an application for summary judgment. If the court is satisfied that the adjudicator has made a valid decision – there should be no enquiry into whether or not the decision is correct – judgment will be given and can then be enforced as any other judgment. The process as described was approved by Sir John Dyson QC, although not in fact followed by the parties, in *Macob Civil Engineering Ltd* v. *Morrison Construction Ltd* (1999). In *Outwing Construction Ltd* v. *H Rendell & Sons Ltd* (1999) Judge Humphrey Lloyd QC went further and approved an application to abridge the normal time periods for summary judgment procedures. The process should take no more that a few weeks and, in the light of *Outwing*, may only take few days.

Alternatively, the Scheme for Construction Contracts (as it applies to England and Wales) provides a somewhat complex procedure borrowed from the Arbitration Act 1996 enabling an application to be made to the courts for enforcement. Sir John Dyson, in the *Macob* case, suggested that this procedure was best reserved for the enforcement of decisions that involved the requirement that one party take some specific action other than payment of money. The Scottish equivalent is rather more simple and enables the successful party to enforce a decision in a very similar way to a Scottish arbitration award.

The original dispute may still be taken to arbitration (or litigation) and the adjudicator's decision may be reversed or otherwise changed.

Adjudication in practice

Contractors have been reluctant to take disputes to adjudication under the new legislation and at the time of writing (April 1999) there has been little experience of the process in either civil engineering or building projects. Many engineers, surveyors, lawyers and others have been trained as adjudicators by the various bodies offering to act as appointors of adjudicators – such as the Institution of Civil Engineers and the Chartered Institute of Arbitrators. These potential adjudicators understand that the task of coming to a decision in a complex claim matter on a civil engineering project within 28 or 42 days, as they may well be obliged to do, is daunting indeed.

Contractors and sub-contractors making justifiable claims should not be afraid of the process of adjudication. If impasse has been reached with regard to a claim and the person making the claim still believes that there is a clear entitlement, adjudication should be considered seriously. The following should be borne in mind:

- The adjudication process is much better suited to dealing with single issues than complex claims comprising many causes of action. If there are several claims to be made, it may be better to refer them one at a time rather than combining them all in one reference as would be appropriate in arbitration.
- The process is designed to produce a very fast decision during

189

the currency of the works, preventing relatively minor grievances from developing into major disputes. The earlier a matter is referred, the more likely it is that this will be achieved.

- The period of 28 days is a very short one. The case to be submitted to the adjudicator should be fully prepared before the timetable starts. It should be expressed in the clearest possible terms.
- Having started the process, there must be a commitment to compliance with the timetable.
- Lawyers are not always necessary but may be able to assist in achieving clarity.

CHAPTER EIGHT
PREPARATION AND NEGOTIATION OF CLAIMS

8.1 Introduction – the basic requirements

From the viewpoint of the claimant, the objective in any 'claim situation' is to obtain the payment to which he believes himself to be entitled, whether under the contract or otherwise. To achieve this objective it is necessary for the claimant to define clearly the basis of his claim, and to produce evidence to support it. It is for the claimant to establish that the claim is legally valid, and he must also substantiate the amount claimed – figures cannot be plucked out of the air. The onus of proof rests on the claimant. The legal maxim 'he who asserts must prove' applies.

In practical terms, the contractor must establish

- the contract provision on which he relies;
- that he has complied with the contractual requirements as to the giving of notices; and
- that the additional cost or expense has been incurred as a direct result of the event relied on.

Claims are not a means of turning a loss-making contract into a profitable one; but investigation of the causes of a loss may sometimes indicate where a claim may lie.

8.1.1 Standard of proof

The standard of proof in civil actions and in arbitration is on 'the balance of probabilities'. This is of course a much lower standard than that required in a criminal case in which the prosecution must establish guilt 'beyond reasonable doubt'. In exercising his independent role in dealing with contractual claims, the engineer should take as his criterion the former standard. Proof on the balance of probabilities means that the evidence in support of the claim is more persuasive than that against it.

8.1.2 'Clean hands'

The claimant must also recognise his contractual obligation to carry out and complete the works in accordance with the terms of the contract and within the time prescribed by it. He should have 'clean hands'. In practical terms this means that the contractor should have complied fully with his contractual obligations; he should not find himself in a position in which he has to seek the engineer's indulgence in respect of substandard work, unwarranted delay, or of some other aspect of his performance while making a claim in connection with that work.

8.1.3 Mitigation

It is a common belief that a claimant is under a general duty at law 'to take all reasonable steps to mitigate the loss consequent on the breach' – which belief originates from the words of Viscount Haldane in *British Westinghouse & Manufacturing Co Ltd* v. *Underground Electric Railways Co of London Ltd* (1912). More recently the true legal position was accurately summarised by Sir John Donaldson MR in *The Solholt* (1983):

> 'A plaintiff is under no duty to mitigate his loss despite the habitual use by lawyers of the phrase "duty to mitigate". He is completely free to act as he judges to be in his best interests. On the other hand, a defendant is not liable for all loss suffered by the consequence of his so acting. A defendant is only liable for such part of the plaintiff's loss as is properly to be regarded as caused by the defendant's breach of duty.'

What this means is that the claimant must act reasonably. Where for example the contractor is delayed for some reason for which he is not responsible, he should consider whether it is sensible and practical to take hired-in plant off hire, and whether he could usefully deploy his own resources elsewhere. In many instances, however, the information needed to assess the economic merits of returning the hired plant to its owner, and so saving hire costs at the expense of additional haulage costs, is not available until after the event. Often the duration of the delay, which may have been caused by lack of information, by an unforeseen obstruction, by an engineer's direction or variation, by unavailability of the site or by the fault of a nominated sub-contractor, is not known until after the delay has ended. It is suggested that where the basic information needed for a rational decision on such matters is not available, the contractor should make his own assessment and decision, and should invite the engineer to approve or to reverse that decision. In that way the danger that the engineer might, with the wisdom of hindsight, be able to reject a claim may be avoided.

Probably the most usual reason why many potentially valid claims are abandoned is the claimant's failure to maintain adequate records, either of events or of cost or of both. Without proper relevant and contemporaneous records, the evidence needed to establish the claim is not available and the claim is doomed.

Where records can be agreed contemporaneously with the engineer a potential source of dispute is eliminated. Failing such agreement, the claimant should himself maintain proper, full and accurate records, should give the engineer an opportunity to check and to agree them as they become available, and should ensure that he has evidence of having done so.

8.2   Recording the facts

8.2.1   Agreeing records

As soon as the contractor becomes aware that he has, or may have, a claim under the contract, he should ensure that full and detailed records relating to the history of the claim are prepared and are maintained throughout the period during which conditions giving rise to the claim remain in existence. These records should be submitted to the engineer for his agreement. Where the engineer refuses or fails to check the contractor's records, that fact should be recorded in a letter to the engineer.

An engineer's refusal to agree records, usually where he does not admit the existence of a claim, does not help his client if the contractor can later establish that his claim is valid. If the claim is ultimately referred to arbitration, the arbitrator can only consider the evidence before him. Where a contractor has maintained

193

accurate and contemporaneous records which the engineer has refused to check or to agree it may be difficult indeed for the engineer to produce convincing evidence to refute the claim. Where, as is sometimes the case, the engineer maintains separate records, a problem will arise if and to the extent that the engineer's records differ from those of the contractor.

In such a situation, the arbitrator can only weigh the evidence before him and decide which is the more convincing. The fact that the contractor had offered to make his own records available to the engineer for inspection, but that the offer had been refused, would weigh heavily in the contractor's favour, assuming of course that there is no evidence to prove that the contractor's records were inaccurate.

8.2.2 Notice

Notice of claim is required under clause 52(4) of the ICE Conditions, and may be required under the particular clause under which the claim is to be made. For example, there are special notice provisions relating to claims under clause 12 and to those for extension of time under clause 44.

There is no general notice provision under the Minor Works Conditions. Clause 3.8 (adverse physical conditions) requires written notice 'as early as practicable' and clause 4.4 (extension of time) requires a written request from the contractor, which may in turn trigger a claim for additional cost under clause 6.1.

Frequently the full extent of the claim, and certainly its monetary value, is not known at the time when the contractor becomes aware of its existence. In such cases, the contractor should give such information as is available with his notice of claim and he should supplement that information from time to time as further details become known.

8.2.3 Content of records

Records should include the date and, where appropriate, the time of the event giving rise to the claim, the action taken to deal with delays, disruption or obstruction of work, other relevant information such as the stage reached in the construction of the works affected, and weather conditions where they are relevant to the claim. It is also essential that there be adequate records of cost or expense involved and the contractor must be able to establish that the additional cost was *caused* by the event that is relied on. Records for the purpose of evaluation of the claim must therefore be maintained. (See section 8.3 below.)

8.2.4 Progress records

These are of particular importance in a delay situation, and a comparison of a critical path programme with a detailed record of site progress is a sensible way of assessing the extent of the employer's responsibility for the delay and his liability for additional costs. Even with the fullest of records of actual progress, it is necessary to show a causal link between the events which form the basis of the claim and the delay.

Clause 14 of the ICE Conditions requires the submission of a programme by the contractor although the programme is not normally a contractual document. The programme as approved by the engineer is in fact a very useful tool for assessing any delay and may provide part of the evidence needed to substantiate a claim for additional cost.

8.2.5 Correspondence

Relevant letters, memoranda and other documentary evidence such as drawings received from the engineer may be of assistance in supporting a claim. It is essential that these should record events as they happen, with suitable reference to the relevant contract clauses.

8.2.6 Minutes of meetings

The minutes of progress or site meetings provide useful supporting evidence and it is essential that any alleged inaccuracies in such records are challenged at the time they are circulated. Failure to do this may be taken to imply acceptance of the minutes as a true record of the proceedings.

8.2.7 Photographs etc.

Photographs and videos showing the actual situation on site are both relevant and admissible, provided that they are clearly marked to indicate the date, location and direction of the view.

8.2.8 Site diaries

These are often useful evidence as to progress and what happened on site provided that they are accurate and reasonably unbiassed. The contractor's agent's site diary should, for example, record complaints as well as progress. A comparison of the RE's diary with that of the agent may reveal very different viewpoints and their attitude in recording the same events.

In *Oldschool* v. *Gleeson (Construction) Ltd* (1976) there was a conflict between the diaries of the consulting engineer and the contractor's site agent, and the judge found the agent's diary entries

'unsatisfactory in the sense that they do not record warnings or complaints when, as I believe, these were given. I cannot believe that the district surveyor would have written [to the contractor recording his assistant's visit to the site and warning against further excavation] if [the assistant] had not in fact given the warning, and yet not a hint of it appears in [the agent's] diary; which, together with other instances, leads me to think that he was not anxious to record criticisms or complaints when they were made, and where such appear in [the consulting engineer's] diary and not in the diary of [the agent] I am bound to say that I have no hesitation in accepting [the consulting engineer's] contemporaneous record as being the accurate one.'

8.3 Recording cost information

The main elements of the cost of construction work are labour, plant, materials, site overheads and head-office overheads. Where work is sub-contracted, the cost to the main contractor is the amount of his payment to the sub-contractor, evidence of which is normally readily available. However, it is often necessary for that amount to be broken down into the same elements as apply to work carried out by the main contractor himself.

8.3.1 Labour costs

Whether or not a claim has been identified, records of labour costs for the whole site are almost invariably available in the form of wages sheets – now usually on computer files. These cover most, but not all, of the costs related to the employment of labour. Other related costs for which records may be required are labour-related insurances, holiday-with-pay schemes, sick-pay schemes, payments in respect of 'wet time', guaranteed minimum wages or bonus payments for periods of inclement weather, industrial training levies, redundancy payment contributions, statutory liabilities, small tools, and protective clothing. Certain items in the wages sheets, such as lodging and subsistence allowances, may need to be apportioned where only part of a week's work is allocated to the claim.

Where the engineer orders that work forming the subject of the claim is to be paid for as daywork, the majority of these cost elements are covered by the standard percentage addition to the 'amount of wages' as defined in the CECA *Schedules of Dayworks* or at the rate entered by the contractor in the contract daywork schedule. In such cases, care must be taken to ensure that there is no duplication – for example, that supervisory staff whose costs form an element of the percentage addition to wages are not included in the list of men for whom labour costs are claimed.

In order that the costs of labour associated with the claim may be determined, it is essential that the men affected by the work or by the delay can be identified. Usually, productive men on site are organised into gangs, which move from one section of the project to another. In that case the identification required may be just the names of the men in each gang and daily records of the section of work on which each gang is engaged.

8.3.2 Plant costs

Where plant for work forming the subject of a claim has to be hired from an external source, or where hired-in plant is subject to delay by a cause for which the employer is responsible, the actual payment to the plant hire company will be the plant cost. In such cases the contractor's claim is limited to that part of the cost that is incurred necessarily. It follows that, subject to what is said below, in an appropriate case he should take the plant off hire.

A difficulty arises where the contractor's own plant is used for work, or where the contractor's own plant becomes unproductive because of some delay for which the employer is responsible. The contractor will usually seek to charge either contract daywork rates or CECA daywork rates for such plant, or rates that he would charge for hire to another contractor, or else the published rates of the contractor's plant-hire associate.

This is a problem area, and it is suggested that the correct approach is to limit the cost to depreciation and maintenance, as was held by the Court of Appeal in *Bernard Sunley & Co Ltd* v. *Cunard White Star Ltd* (1940), unless, of course, the contractor can establish that he normally hires out plant to other contractors when he is not using it himself. This is a matter of evidence and the engineer should require proof that, had the plant not been required for the work in question, or had it not been tied up unproductively on the site, it would probably have been hired out to another contractor.

Where the contractor's own plant becomes unproductive for some reason for which the employer is responsible, a lower 'standing' charge per hour or per day should be substituted for the 'working' charge, which takes account of fuel and of wear and tear. A difficulty often arises when deciding whether to return plant to the hire company or the depot when a delay occurs. While the costs of transporting the plant to and from the site and the daily or weekly hire costs are known, the duration of the delay, and hence the unproductive hire cost, is usually unknown. In coming to his decision the contractor can only try, as an experienced contractor, to assess the likely delay.

It is suggested, however, that the engineer should be notified of the contractor's decision and given the factual and estimated data upon which it is based. The engineer would then be able, if he thought the contractor's decision incorrect, to overrule it – but he would be unable to contend at a later date that the contractor should have foreseen that the delay was likely to last a longer or a shorter period as the case may be.

8.3.3 Material costs

The increased cost of materials does not usually form a major element of a civil engineering claim but there are certain types of claim in which material costs may need to be recorded and recovered. Examples are materials needed for varied works, temporary works materials such as steel sheet piling needed to deal with subsoil conditions giving rise to a claim under clause 12 of the

ICE Conditions or clause 3.8 of the Minor Works Conditions, and materials needed to rebuild defective works for which the employer is responsible. Evidence of the cost of materials used for such purposes is in most cases readily available in the form of the supplier's invoices.

8.3.4 Site overheads

The main elements of site overheads, which are included in the definition of 'cost' under clause 1(5) of the ICE Conditions and clause 1.3 of the Minor Works Conditions, are set out in Chapter 3, section 3.1.1. Because site overheads form an element of almost every claim, it is essential that the contractor should identify and maintain a record of such costs throughout the construction period. The principles to be used in recording such costs are those given above, but it is necessary to distinguish between 'lump-sum' and 'time-related' costs.

'Lump-sum' costs include

- the cost of hauling to and from the site and erecting site offices, workshops, stores and other buildings;
- the cost of constructing site access roads;
- the initial survey and setting-out if carried out by someone other than the site engineer; and
- installation charges for water supply, sewerage, electricity and telephone.

'Time-related' costs include

- wages and salaries of all 'non-productive' staff;
- rental charges for offices and buildings;
- hire charges for general plant such as cranes and transport not related to any specific item of work;
- heating, lighting, cleaning and telephone rental and call charges
- stationery and postage; and
- all maintenance costs such as the cost of maintaining the site access roads and any other temporary works (such as scaffolding) required for the construction works as a whole.

Some site overheads are not necessarily present throughout the whole period of delay.

8.3.5 Off-site overheads

The method of evaluating off-site overheads has been discussed in section 3.1.2. Whatever system of evaluation is to be used, the factual records are generally those forming part of the contractor's main accounts, or the estimator's pricing notes for the contract from which the claim arises. Off-site overhead costs are clearly a permissible head of claim, and in general are concerned with the consequences of such resources being tied up and unable to generate an overhead contribution elsewhere.

The Canadian case of *Ellis-Don* v. *The Parking Authority of Toronto* (1978) is of interest in this respect. The contractors had been delayed by just over 32 weeks, of which 17.5 weeks was due to the employer's failure to obtain an excavation permit and to the consequential carrying out of concrete pouring operations during the winter. Mr Justice O'Leary allowed the contractor's claim for damages for breach of contract, including a weekly sum in respect of the cost of overheads by reference to the percentage figure built into the tender.

A key passage in the judgment emphasises that the tender figure effectively related to actual overhead costs despite the fact that a formula approach was adopted. The judge said:

> 'There is ample evidence tendered before me that the men and resources tied up for an extra 17.5 weeks through the fault of the defendant could have been readily employed elsewhere and could have continued to earn overheads ... for the plaintiff ...'

8.4 Compliance with and confirmation of engineer's instructions

Many claims originate from an instruction of the engineer or of his representative to vary the work or its sequence or method of construction, or to suspend part or all of the works. The instruction forms best evidence of the event giving rise to the claim, and hence it should be carefully filed so that it may be produced when necessary.

If the instruction is given orally and is not confirmed in writing the contractor should himself confirm it under the procedure set out in clause 2(6)(b), where necessary clarifying the effect of the instruction. Thus, for example, a contractor who receives a vague request to defer part of the work should ascertain whether it is in fact merely a request – which the contractor is under no obligation to meet – or an instruction. If it is an instruction it is presumed to have been given under clause 40(1) and, when confirming or acknowledging receipt of the instruction, the contractor should refer to clause 40(1).

While the contractor has a duty to complete the works, and may not therefore rely upon being given instructions as to how to overcome difficulties (see, for example, clause 12), his own proposals as to how to deal with those difficulties may be approved or rejected by the engineer who retains overall control over both the works themselves and the methods of construction.

By stating clearly, and as soon as possible after the difficulty has arisen, how he intends to deal with it, the contractor avoids a possible allegation that he has failed to mitigate his loss. In particular, he avoids the risk that the engineer may, with the wisdom of hindsight, claim that some method other than the contractor's proposal should have been adopted. By clarifying his proposals, the contractor invites approval or variation of those proposals and avoids a charge that some other method should have been used which would have reduced the amount of the claim.

8.5 Identification of justification for a claim

As soon as possible after the event giving rise to a claim, the contractor should identify the clause or clauses under which the claim is to be made, or any other express or implied term of the contract to be relied upon as the basis of the claim. The notice and any other requirements of the clause or clauses should be observed.

In many cases, the event may give rise to claims under a number of complementary or alternative clauses, or perhaps under a combination of express and implied terms. Express terms should, of course, be relied upon in preference to implied terms, since the existence of the latter is open to challenge. However, where alternatives are available, and in particular where differences as to findings of fact may affect the relevance of express or implied terms, the safest procedure is to keep all possibly valid bases of claim open. This suggests that the notice requirements of each of the possible clauses upon which the claim may be based should be complied with until such time as it becomes clear that any of those bases is no longer valid. In *Fairclough Building* v. *Vale of Belvoir Superstore Ltd* (1990) it was held by Judge Thayne Forbes QC, allowing an appeal against an arbitrator's award, that a contractor could still pursue a claim for damages for breach of contract when a claim based on the same facts under clause 24 of the JCT Form of Contract had failed.

As we have seen, clause 52(4) of the ICE Conditions is a general notice requirement covering claims for 'any additional payment' (other than ordered variations or re-rating) 'pursuant to any Clause of these Conditions'. Other contract provisions may impose other requirements on the contractor. Clause 7, which deals with further drawings and instructions, is an example.

In order to establish a claim under clause 7(4) the contractor must, in addition to complying with the provisions of clause 52(4), show the following:

- That he gave adequate written notice to the engineer of the drawing or specification required.
- That he applied for the information on a specified date.
- That the engineer failed to issue the information at a reasonable time, i.e. the time when he should have issued it and, if appropriate, the date when he did issue it.
- That he suffered delay or incurred reasonable cost as a direct result of the engineer's default.

8.6 Evaluation and submission of claims

The principles upon which evaluation of the two main categories of claim – variations and delays – are based have been covered in Chapters 4 and 5. Detailed evaluation should commence as soon as the information needed becomes available in order that the claim may be included in interim valuations as soon as possible. This is desirable for several reasons:

- The inclusion of a claim – which may initially be contentious in principle – in an interim valuation draws attention to the contractor's intention to pursue that claim. It provides an opportunity for the engineer to challenge its validity, either in principle or in the method used for its evaluation. If it is found that there is some shortcoming in either aspect of the claim as submitted, early submission provides an opportunity to remedy the defect and to present information in a form acceptable to the engineer in subsequent valuations.
- The interim valuations help the engineer to foresee, and the

employer to budget for, the likely amount of the claim, and enable a variation order to be issued if necessary so as to keep the project within the employer's budget.

- Inclusion of the claim in interim valuations is advantageous for the contractor because it provides a commencement date for calculation of interest should the claim be disputed. In addition to the contractual right to interest on late payments provided by clause 60(7), section 49 of the Arbitration Act 1996 empowers an arbitrator at his discretion to award simple or compound interest on money awarded, and on money claimed in arbitral proceedings but paid before the date of the award. In doing so, the arbitrator would have regard to the date upon which the money should have been paid. That date in general is the date upon which payment would have been made had the claim been admitted by the engineer, and clearly must take account of the date by which the contractor had provided information in sufficient detail to permit proper evaluation.

Where, as sometimes happens, the engineer objects to the contractor's inclusion in interim valuations claims that have not been agreed, the contractor should maintain his right to do so. If he does not he may find his entitlement to payment reduced or challenged.

Many of the 'claims' clauses of the ICE Conditions refer to payment in accordance with Clause 60. The intention is to deal with such claims in interim valuations and the position is the same under the Minor Works form. If the reason underlying the engineer's objection to the inclusion of claims that have not been agreed is administrative convenience, the matter may be resolved by the contractor submitting, with his monthly valuation, a separate schedule of claims which have not been agreed.

8.7 The global approach and its limitations

8.7.1 Multiple and interacting causes of delay

It is not uncommon for construction work to be delayed by a large number of variations and other instructions of the engineer. Claims are in such cases presented in respect of each individual instruction, but is the engineer justified in insisting that the cause of each element of the overall delay must be identified, evaluated in terms of days or weeks of delay, and proved?

The well-known case of *J Crosby & Sons Ltd* v. *Portland UDC* (1967) provides authority for a negative answer to that question. Mr Justice Donaldson (as he then was) accepted the validity of the contractor's argument that where a claim depends on 'an extremely complex interaction in the consequences of various denials, suspensions and variations, it may well be difficult or even impossible to make an accurate apportionment of the total extra cost between the several causative events'.

In this situation, therefore, it is not necessary for the contractor to link each item with each separate cause of claim. This principle applies to claims both for time and for money. In *Crosby*, the contractor made a general claim for delay and disorganisation. The completion of the contract had been delayed by 46 weeks by a combination of factors, some of which entitled the contractor to extension of time and/or cost while others did not. The arbitrator's award included a statement which was quoted by Mr Justice Donaldson, as follows:

'The result in terms of delay and disorganisation, of each of the matters ... was a continuing one. As each matter occurred its consequences were added to the cumulative consequences of the matters which had preceded it. The delay and disorganisation which ultimately resulted was cumulative and attributable to the combined effect of all of these matters. It is, therefore, impracticable if not impossible to assess the additional expense caused by delay and disorganisation due to any one of these matters in isolation from other matters.'

The arbitrator's award of a lump sum by way of compensation in respect of 31 weeks of the overall delay was upheld.

Mr Justice Donaldson made two important provisos to this global approach:

- There must be no duplication.
- The sum awarded must exclude profit, if profit is irrecoverable under one or more of the heads of claim.

The 'global approach' to claims was confirmed by Mr Justice Vinelott in the building contract case of *London Borough of Merton* v. *Stanley Hugh Leach Ltd* (1985), where he followed Mr Justice Donaldson. He made the point, however, that it is implicit in the reasoning of *Crosby*:

'[F]irst, that a rolled-up award can only be made in a case where the loss or expense attributable to each head of claim cannot in reality be separated, and secondly that a rolled-up award can only be made where apart from that practical impossibility the conditions which have to be satisfied before an award can be made have been satisfied in relation to each head of claim.'

It is submitted that this remains the position.

8.7.2 Complexity increasing the delay

In many cases it may be argued that the multiplicity of instructions was such as to cause overall delay greater than the sum of the delays attributable to the individual causes.

Where a claim is founded upon the allegation that a multiplicity of variations, instructions, or other actions by the engineer caused a delay in excess of that resulting from each individual element, the first step is to ascertain the overall period of delay to the works, and then to break down that period into its constituent elements:

(1) Delays for which the contractor is responsible under the terms of the contract, and in respect of which no extension of time is due (for example, delays caused by inadequacy of labour, plant or materials).

(2) Delays for which the contractor is financially responsible, but in respect of which he is entitled to an extension of time (for example, delays caused by exceptionally inclement weather).

(3) Delays that can be specifically identified for which the employer is responsible (for example, delays caused by variations, late issue of drawings, delayed access to the site).

(4) The balance of the delay, having deducted (1), (2) and (3) from the total period.

Where it is possible to evaluate the delaying effects of individual causes, then clearly this should be done, leaving only the balance of the delays (4) for which the employer is alleged to be responsible to be covered by the 'global approach'.

8.8 Negotiating settlements

Neither the ICE Conditions nor the Minor Works form supposes that claims will necessarily be settled by negotiation and accordingly both forms make provision for ultimate determination by a third party: namely an arbitrator. Both forms do however seek to promote settlement by agreement.

Clause 52(1) of the ICE Conditions provides expressly for 'consultation with the Contractor' before the engineer 'ascertains' the value of a variation, while clause 6.1 of the Minor Works form makes similar provision in respect of additional works or additional costs arising from delay or disruption. Similarly clause 3.8 of the Minor Works form implies that there shall be consultation between the contractor and the engineer in respect of the cost of dealing with adverse physical conditions etc, before the engineer determines a sum to be paid.

It is to be hoped and indeed expected that in many cases the consultation process will lead to an understanding by the engineer of the basis on which the contractor has built up his proposed rates, and hence to agreement; and it is perhaps surprising that provision for consultation is not made in other potentially contentious 'claims' clauses of the ICE Conditions and of the Minor Works form.

However a structured procedure for conciliation before resorting to arbitration, first introduced into the Minor Works form, has now been included in the Sixth Edition.

The use of the machinery provided in the forms confers many benefits on the contractor, not the least of which is that he is entitled to have amounts found due included in interim payments – provided, of course, that he has adhered to the machinery set out in the contract.

In practice, many claims are, wisely, settled by negotiation between the contractor and the engineer. Negotiated settlements are a desirable objective for both parties to the contract. Failure to negotiate a settlement may result in abandonment of a valid claim if the contractor does not wish to embark upon the uncertainties and hazards of arbitration or litigation, both of which involve expenditure of time and money for which even an award of costs to a successful claimant may not provide full recompense. Failure to negotiate often leads to delay and ill-will. Furthermore the employer may have to pay more under an arbitrator's award than would have been acceptable in an amicable settlement; and in addition an award of costs.

Most claims are settled by negotiation. Even if a dispute is referred to arbitration, it is likely that a settlement will be agreed before the start of the hearing. It should be recognised that commencement of arbitration does not close the door to a negotiated settlement. On the contrary, a reference to arbitration is likely to engender a sense of urgency into any negotiations between the parties.

However, both parties should beware of the dangers of abdicating their functions to lawyers once arbitration proceedings have commenced. Managers and quantity surveyors sometimes believe that their duties are terminated when their legal advisers commence proceedings – files and other records are passed to the lawyers, and negotiations between technical staff come to an end. Lawyers and technical staff – engineers, quantity surveyors, and architects – should see their functions as being complementary both before and during the arbitration; and both should have free access to all relevant records in order that negotiations may continue.

Under clause 66(5) of the Sixth Edition and clause 11.3 of the Minor Works form the option of conciliation provides a sensible and practicable means whereby disputes (including disputed claims) may sometimes be settled with a minimum of delay by obtaining an independent recommendation as to how the dispute should be settled. Such settlements are however achieved only where both parties have a genuine wish for negotiations to succeed.

During negotiations the contractor should be prepared to provide all written evidence that the engineer may reasonably require to establish the validity of the claim. If the claim has ultimately to be referred to arbitration, the respondent will have the right to demand such evidence. This applies in particular to the contractor's pricing notes, which are often considered by the contractor to be confidential but which would nevertheless have to be produced at the stage of 'disclosure of documents' in an arbitration. The contractor may, however, require that any confidential information submitted in support of a claim is used only for that purpose and is not disclosed to third parties. Similarly, the contractor should be prepared to discuss his claims and, if requested to do so by the engineer, arrange for the relevant personnel to attend meetings with the engineer in support of the claim.

Additionally, the contractor should consider carefully and determine an acceptable 'settlement figure' for the claim. He should recognise that his objective is to obtain payment of the sums to which he considers himself entitled and which he can substantiate. An exaggerated and unsupported claim is doomed to failure. There is therefore nothing to gain by the submission of inflated claims. However, most contractors' claims tend to be optimistic. This is perhaps understandable, because any doubt about the validity of a claim will initially be decided by the contractor in his own favour. He will also apply the most favourable method of evaluation where alternatives exist.

From the outset the contractor should have made a realistic assessment of the value of the claim for use as a datum in considering any offer of settlement. One method of assessment is to estimate a factor representing the likelihood of success of each item of the claim and to multiply that factor by the likely, as distinguished from the optimistic, evaluation of that item. Thus, for example, in a £100 000 claim for delay there may be a 50% chance of success on the question of liability, and a realistic evaluation of the claim, if it succeeds, of £80 000. Hence the value of the claim to the contractor is assessed at £40 000 for purposes of settlement.

Of course there is a considerable element of gambling in using such methods of evaluation but where the claim contains many items it is to be hoped that the overall evaluation may be reasonably accurate.

Once the contractor has provided the engineer with all the supporting information he requires, and it has been considered and evaluated, the contractor may expect to receive the engineer's reply. This will usually include the engineer's assessment of the value of the claim. Outright rejection by the engineer of a claim is not unusual, especially in claims based on clause 12, where the engineer states that the physical conditions of artificial obstructions were such as could reasonably have been foreseen by an experienced contractor, or clause 7, where the engineer denies that any delay was occasioned by his allegedly late issue of drawings.

If a claim is rejected by the engineer, the reasons for the rejection should be considered. It is open to the contractor to seek to persuade the engineer of the validity of his claim by further written or oral submissions. The wise engineer will allow a reasonable opportunity for the contractor to do so, recognising that provision of an opportunity to submit a case is one of the basic requirements of natural justice. When a stage is reached at which the contractor considers that nothing is to be gained by further written or oral submissions, he must decide whether to accept whatever the engineer may have offered in settlement, to invoke clause 66 under the ICE Conditions, or to seek conciliation if appropriate.

If the engineer rejects the claim in its entirety, the contractor should realise that a reference to arbitration will incur a risk that he may have to bear the costs of both the parties and also the arbitrator should the arbitrator uphold the engineer's decision. The amount at risk is clearly less under the conciliation procedure. On the other hand, an award to the contractor of even a proportion of the claims referred to the arbitrator should carry with it the contractor's own costs – provided that they have been incurred necessarily – and protection against the employer's and the arbitrator's costs.

It follows that if the contractor, having taken legal or technical advice as necessary, is confident that at least one major item in his claim has been rejected unfairly, he may be similarly confident of success in a reference to arbitration. But his claim need not then be confined to the one item. Other items of less predictable outcome may be included without incurring undue risk of an adverse award of costs. This is because, in general, arbitrators should not apportion costs even where only a proportion of a claim succeeds: *Channel Island Ferries Ltd* v. *Cenargo Navigation Ltd (The Rozel)* (1994). The fact that the claimant has had to invoke arbitration in order to obtain money to which he has shown himself to be entitled should also entitle him to an award of costs.

There is, however, an important proviso. The time of the arbitrator and the other party should not have been wasted in the pursuit of obviously invalid claims. Where an item in the claim comes within this category and the additional time spent can be identified it is to be expected that the arbitrator, notwithstanding the claimant's success in his other claims, will award costs against the claimant in respect of the time so wasted.

Where the contractor's submission to the engineer results in an offer in settlement, the contractor should consider carefully the significance of that offer in relation to his own realistic assessment of the value of the claim. If the offer meets or exceeds the contractor's own realistic assessment (which may include allowances for any interest that may have accrued and for any costs that may at the time of the offer have been incurred *necessarily)*, then clearly he should accept the offer.

Where the offer is less than the contractor's realistic assessment then, provided he is confident of the accuracy of his assessment, the contractor should be similarly confident in taking matters further. However, there may be a considerable element of uncertainty in the outcome of any reference to arbitration. An award even marginally below the amount of the offer received before the arbitration is commenced will put the contractor at risk in respect of all costs incurred after the date of the offer. Where the claim is of modest value, such an outcome could result in the contractor having to bear costs in excess of the sum awarded to him. Hence in those circumstances it may be prudent for the contractor to accept an offer below his own realistic evaluation as a safeguard against an adverse award of costs in an arbitration.

Another possible course of action upon receipt of a global offer in settlement is to request the engineer to define the sums awarded against each item of the claim. If the sums awarded against any items are acceptable to the contractor he may indicate this to the engineer and request certification of the sums offered, so eliminating those items from the list of disputed claims. By so doing, the matters remaining in contention and which may have to be referred to arbitration can be reduced, with a consequent reduction in the costs of the arbitration. Furthermore such a course of action may result in renewed attempts to negotiate settlement of the remaining items.

Negotiations for a settlement could, in some circumstances, have a possible effect on the contractor's entitlement. Clause 52(4)(b) requires the contractor to give notice of intention to claim 'as soon as may be reasonable and in any event within 28 days after the happening of the events giving rise to the claim' and, while case law suggests that this may be interpreted generously, this should not be relied on.

Tersons Ltd v. *Stevenage Development Corporation* (1963) – a case on the provisions of the Second Edition of the ICE Conditions – which is usually cited in this connection is, in fact, of very little assistance in interpreting clause 52 of the Sixth Edition of the ICE Conditions. Clearly, the contractor should give notice as soon as possible after the event, even if only in general terms. If negotiations are embarked upon, it is suggested that these should be on the basis that they are without prejudice to the contractor's strict contractual entitlement. This would meet any subsequent argument that the contractor had not complied with the provisions of clause 52(4), thereby limiting his claim by virtue of clause 52(4)(e).

There appears to be no case directly in point in a civil engineering context. Once again, however, there is guidance from the related (though more litigious) field of building contracts. In *Rees & Kirby Ltd* v. *Swansea City Council* (1985), the Court of Appeal was concerned with the provisions of the JCT 63 Standard Form of Building Contract, the claims clauses of which required the contractor to give notice in writing within a reasonable time of the loss and/or expense having been incurred. The formal application was, in fact, made several years after the event, but the employer was precluded from objecting to the late claim because by his conduct he had led the contractor to believe that a negotiated settlement was likely, and so he could not rely on his strict legal rights.

This is known as the principle of 'promissory estoppel'. It arises where the parties enter 'upon a course of negotiation which has the effect of leading one of the parties to suppose that the strict rights arising under the contract will not be enforced' (Lord Chancellor Cairns in *Hughes* v. *Metropolitan Railway Co* (1877)).

In *Rees & Kirby*, as soon as the contractors raised the question of a claim under the contract, the parties entered into negotiations for an *ex gratia* settlement – negotiations which lasted a number of years. Even when the negotiations came to an end, the employers raised no objection to the contractors instructing their own quantity surveyor to prepare a claim, nor did they notify the contractors that they intended to resume their right under the contract to a written application. These points, amongst others, precluded the employers from objecting that the contractors' claim was out of time.

Despite this guidance, it is as well for the position to be made clear from the outset of any negotiations and, if necessary, the disputes procedure under the contract should be invoked as a precaution.

8.9   Example of a claims submission under ICE Sixth Edition: I

Note: This is one possible form of submission and the assumptions on which it is based would in practice need to be substantiated by evidence. It is not an example to be followed slavishly.

Sewerlayers Ltd v. Wessex District Council

THE CONTRACT

(1)   The contract comprises a tender submitted on 19 January 1996 by Sewerlayers Ltd ('the Contractor') to the Wessex District Council ('the Employer') for sewers, manholes, service connections, two pumping stations and two rising mains required for a regional sewerage scheme to serve the villages of Buckholme and Westerview in the Wessex District, and the Employer's implied acceptance thereof in that the Employer's Engineer, Thomas Telford ('the Engineer'), instructed the Contractor to commence work on 1 April 1996. The tender total, being the total of the bill of quantities, is

£3.343 million. The time for completion stated in the Appendix to Form of Tender is 78 weeks.

(2) The contract incorporates the ICE Conditions of Contract, Sixth Edition, as revised in July 1993. Other documents incorporated in the contract are the drawings, specification and bill of quantities, and the Contractor's letter dated 19 January 1996 wherein those other documents are listed.

NARRATIVE

(3) Work on site was commenced in accordance with the Engineer's instruction on 1 April 1996. On 8 April 1996 the Contractor submitted to the Engineer a programme of the works pursuant to clause 14(1) of the conditions of contract, which programme provided *inter alia* for work on the pumping station at Buckholme to be commenced on 3 June 1996 and to be completed by 31 October 1996. Thereafter the pumping station at Westerview was programmed to be constructed, in a construction period of 20 working weeks.

No comment on the programme was made by the Engineer, and accordingly the programme was deemed to have been accepted, pursuant to clause 14(2) of the conditions of contract, on 29 April 1996.

(4) By 17 May 1996 the access road to Buckholme pumping station had been constructed and steel sheet piling, a piling hammer and crane, together with ancillary equipment, had been assembled on site in preparation for construction of the temporary cofferdam needed for the deep excavation; all in accordance with the Contractor's programme.

(5) On 24 May 1996 the Engineer telephoned the Contractor's agent, Mr Brassey, requesting him to defer work on Buckholme pumping station because of impending alterations to the design, and to start work instead on the other pumping station at Westerview. Mr Brassey replied that that alteration to the sequence of construction would result in substantially increased costs and that he was unwilling to comply unless instructed to do so. The Engineer thereupon instructed Mr Brassey, orally, to defer work on Buckholme pumping station. The Contractor confirmed receipt of the Engineer's oral instruction by letter to the Engineer dated 24 May 1996.

(6)   In that letter the Contractor pointed out the probable consequences of the instruction as follows:

> 1.   The Contractor had intended to construct the two pumping stations consecutively because they both required the same type of equipment, which could therefore be used twice. Buckholme pumping station had been programmed for construction before Westerview because it involved the deeper excavation, which determined the length of the steel sheet piling needed for the temporary cofferdam to support the sides of the excavation. Because of the nature of the subsoil as indicated by the borehole logs it was likely that driving conditions would be hard, and that the steel sheet piling would sustain damage during driving. By using the piles at the deeper pumping station first it would be possible to burn off any damaged ends after the first use and still have piling of sufficient length for use in the cofferdam required in the construction of Westerview pumping station. The change ordered by the Engineer could result in a need to buy additional piling for Buckholme pumping station since the piling already on site would have to be used initially for the shallower pumping station if further delay was to be avoided, and might thereafter be of inadequate length, after burning off damaged ends, for use at Buckholme pumping station.
>
> 2.   Secondly it had been intended to construct Buckholme pumping station between the months of June to October 1996 inclusive because the pumping station is situated on low ground and close to a river which is often in flood during the winter months. Any substantial deferment was likely to throw the construction of Buckholme pumping station into the winter months and hence to incur the risk of flooding.

(7)   In compliance with the Engineer's instruction steel sheet piling, pile driving equipment and the crane were moved to the site of the Westerview pumping station as soon as an access road thereto had been constructed. Thereafter construction of the pumping station proceeded without incident except that considerable damage was caused to the steel sheet piling during driving, as had been expected. Work on Westerview pumping station was commenced on 20 June 1996 and was completed on 31 October 1996.

(8)   Meanwhile the Engineer had in a letter dated 4 October 1996 stated that there was now no intention to alter the design of Buckholme

pumping station and that work on that pumping station could therefore proceed in accordance with the drawings already issued.

(9)   After having extracted steel sheet piles from the Westerview pumping station cofferdam it was found that 30 of the 56 piles used had been damaged by hard driving to such an extent that after burning off the damaged parts they were of insufficient length for use in constructing the Buckholme pumping station cofferdam. Additional piles were ordered immediately but were not available on site until 6 December 1996.

(10)   Driving of the original undamaged piles and the additional piles for the Buckholme cofferdam was commenced on 15 November 1996 and was completed by 18 December 1996. Excavation within the cofferdam was commenced after the Christmas/New Year holiday and was completed by 22 January 1997, by which date timbering inside the cofferdam had been partially fixed but had not been secured against flotation.

(11)   On 24 January 1997 heavy rainfall resulted in the river overflowing its banks and rising to a level approximately 1 metre above the top of the cofferdam, flooding that cofferdam and causing the timbers to float to the surface. In consequence the sides of the cofferdam collapsed causing extensive damage to the steel sheet piling.

(12)   As soon as weather conditions permitted a heavier crane and a pile extraction hammer were brought to the site and the damaged steel sheet piling was removed.

(13)   Meanwhile new steel sheet piling had been ordered as soon as it became clear that the original piling was so badly damaged as to be unusable, and it was delivered to the site on 26 February 1997. Construction of a new cofferdam was however delayed by the need to drive steel sheet piling through the irregular ground surface resulting from the collapse and using a crane capable of lifting the piling hammer from a greater radius because the crane was not then able to stand close to the cofferdam as originally intended. Work on the construction of the new cofferdam was not completed until 11 April 1997, and excavation was completed on 25 April 1997.

(14)   By letter dated 3 February 1997 the Contractor gave notice under clause 44 of the contract that delays had been caused as a result of the events outlined in paragraphs (5) to (13) above, and requesting an extension of time of 30 weeks.

(15)   The Engineer failed to reply to that letter until 11 April 1997 when, in response to a telephoned request from the Contractor to grant the extension applied for, the Engineer refused any extension stating that the delay had been caused by the inadequate design of the cofferdam timbering, for which the contractor was responsible.

(16)   Construction of the Buckholme pumping station was further delayed by late issue of reinforcement detail drawings, of which notice had been given by the Contractor to the Engineer that such drawings were required no later than 3 January 1997. This was necessary because the Contractor intended to obtain reinforcement ready cut and bent from the supplier, who quoted a delivery period of 8 weeks from receipt of full details and schedules. Reinforcement detail drawings and schedules were not in fact received by the Contractor until 4 April 1997; and although the Contractor immediately ordered the reinforcement from his supplier it was not delivered to the site until 6 June 1997.

(17)   Thereafter construction work proceeded in accordance with a revised programme submitted to the Engineer by the Contractor on 10 June 1997, except that completion of the pumping station was further delayed by the late delivery of pumps from the nominated sub-contractor Fulflow Ltd, against whom the Contractor is pursuing claims in respect of delay and consequential cost.

(18)   Reinforced concrete work at Buckholme pumping station was completed on 30 September 1997. Building work on the superstructure continued thereafter until 25 November 1997, and pump installation until 16 January 1998.

(19)   Making good and internal decoration after the completion of the pump installation took a further two weeks. The Contractor applied for and was granted a certificate of completion which was effective from 9 February 1998.

(20)   The Engineer failed to comply with the requirements of clause 44(3) of the contract in that he failed to notify the Contractor of his assessment of the extension of the time for completion to which the Contractor was entitled on the due date for completion which, in the absence of any extension, remained 29 September 1997.

(21)   On 6 February 1998 the Engineer wrote to the Contractor stating that completion of the works was not achieved until 19 weeks after the due date for completion and that accordingly he had advised the Employer that he was entitled to deduct liquidated damages at the

rate of £6000 per week for 19 weeks, making a total deduction of £114 000, which sum has been withheld from sums otherwise due to the Contractor.

THE CLAIMS

(22)   The Contractor claims the following:

(1)   Pursuant to clause 40(1) of the contract, a nett extension of time of 12 weeks in respect of delays resulting from the Engineer's oral instruction of 24 May 1996 confirmed by the Contractor's letter of 24 May 1996, and consequential effects thereof, to suspend progress on the Buckholme pumping station, together with the extra costs incurred by the Contractor in giving effect to the Engineer's instruction; of which site costs amount to £97 760, particulars of which are given in Appendix 1 hereto.

(2)   Pursuant to clause 7(4) of the contract, an extension of time of 7 weeks in respect of the late issue of reinforcement details of Buckholme pumping station together with extra costs incurred by reason of the said delay; of which site costs amount to £38 570, particulars of which are given in Appendix 2 hereto.

(3)   Pursuant to clause 47(5) of the contract, reimbursement of the sum of £114,000 deducted by the Employer from payments otherwise due to the Contractor as liquidated damages in respect of late completion of the works.

(4)   Off-site overheads in respect of the total period of delay for which the Employer is responsible, namely 19 weeks, in the sum of £38 411, particulars of which are given in Appendix 3 hereto.

(5)   Pursuant to clause 60(7) of the contract, interest on the amounts due to the Contractor in the sum of £37 178, particulars of which are given in Appendix 4 hereto.

(23)   Summary of claims:

(1)	Under clause 40(1)	(Appendix 1)	97 760
(2)	Under clause 7(4)	(Appendix 2)	38 570
(3)	Under clause 47(5)		114 000
(4)	Off-site overheads	(Appendix 3)	38 411
(5)	Under clause 60(7)	(Appendix 4)	37 178

Total   £325 919

APPENDICES

(All of which will be supported by documentary and oral evidence.)

Appendix 1: Costs arising from Engineer's suspension order: clause 40(1)

Labour	Rate/man-week	Man-weeks	Amount
1 No. Ganger	340	12	4 080
2 No. Plant operators	310	24	7 440
4 No. Labourers	290	48	13 920
Total labour costs			£25 440

Plant:	Rate/week	Weeks	Amount
Crane, tracked 20 t	1000	12	12 000
Pile extractor 1.5 t	120	6	720
Piling hammer 2.0 t diesel	700	8	5 600
Pump, 102 mm diesel	70	12	840
Total plant costs			£19 160

Materials:	Rate	Quantity	Amount
Steel sheet piling	540/t	34 t	18 360
Timbers	300/m^3	12 m^3	3 600
Total material costs			£21 960

Site Overheads:	Rate/week	Weeks	Amount
Agent (including car)	380	12	4 560
Site engineer	340	12	4 080
General foreman	360	12	4 320
Secretary/clerk	260	12	3 120
Tea boy/cleaner	240	12	2 880
Office incl services	190	12	2 280
Site workshop/stores	300	12	3 600
Contract insurances	340	12	4 080
Performance bond	190	12	2 280
Total weekly site overheads	2600	12	£31 200

Summary

Labour	25 440
Plant	19 160
Materials	21 960
Site overheads	31 200

Total of site costs: claim under clause 40(1)	**£97 760**

Appendix 2: Costs arising from late issue of drawings: clause 7(4)

(Drawings required for construction of the reinforced concrete substructure of Buckholme pumping station were issued 13 weeks late. The nett delay to the works was 7 weeks.)

Labour	Rate/man-week	Man-weeks	Amount
1 No. Ganger	340	7	2 380
1 No. Plant operator	310	7	2 170
2 No. Craftsmen	310	14	4 340
3 No. Labourers	290	21	6 090

Total labour costs	£14 980

Plant:	Rate/week	Weeks	Amount
Crane, tracked 5 t	700	7	4 900
Pump, 102 mm diesel	70	7	490

Total plant costs		£5 390
Site overheads (details as Appendix 1)	7	£18 200

Summary

Labour	14 980
Plant	5 390
Site overheads	18 200

Total of site costs: claim arising under clause 7(4)	**£38 570**

Appendix 3: Off-site overheads: clause 1(5)

Off-site overheads in respect of a total delay of 19 weeks: using adjusted Emden formula: (see Chapter 3, section 3.1.2).

Contract value:	£3 343 000
HO overheads percentage (including 1% profit):	6%
Total of site costs:	£3 343 000 × 100/106 = £3 153 774
Offsite overheads claim:	5% × 3 153 744 × 19/78 = £38 411

Appendix 4: Interest (calculated up to 31 December 1998)*

Item of Claim	Principal	When due	CIF1	CIF2	Interest
Clause 40(1)	97 760	30.11.96	2.0772553	2.4959891	19 706
Clause 7(3)	38 570	31.07.97	2.1929071	2.4959891	5 331
Clause 47(5)	114 000	28.02.98	2.3118357	2.4959891	9 081
Off-site O/H	38 411	28.02.98	2.3118357	2.4959891	3 060

Interest claim					**£37 178**

* Interest is calculated at 2% above base rate, compounded monthly, pursuant to clause 60(7) of the contract. Compound Interest Factors CIF1 and CIF2 are taken from *Arbitration Practice in Construction Contracts* (4th edn) by Douglas Stephenson (Blackwell Science).

8.10 Example of claims submission under ICE Sixth Edition: II

Note: This is another possible form of submission and the facts on which it is based would in practice need to be substantiated by evidence. As in the first example, it should not be taken to be the only form of presentation of a claim.

Netherlands Dredging Company (GB) Ltd v. Wetsea Water Authority

THE CONTRACT

(1)   The Contractor is a wholly owned subsidiary of NV Amsterdam Dredging of Rotterdam. The Employer is the statutory water authority, responsible for water supply and sewerage within an area covering some 3500 square miles in north-east England.

(2)   By letter dated 28 February 1997 the Employer, through the Engineer Mr I K Brunel, invited the Contractor to tender for the construction of a sea outfall sewer in heading and for ancillary works at Wetsea, Yorkshire, in accordance with the ICE Conditions of Contract, Sixth Edition. The contract documents included a specification, a bill of quantities, tender drawings and site investigation report, as listed in the Engineer's letter of invitation.

(3)   By letter dated 29 March 1997 the Contractor submitted a tender based on the contract conditions, drawings, bill of quantities and

specification. He also submitted an alternative tender for construction of the outfall by dredging at the seaward end and by open-cut tidal excavation at the shore end of the outfall.

(4)   The Contractor's alternative offer was in the form of a lump sum price for providing and laying the outfall sewer. The amount of the alternative tender was £5 850 000, which was some £900 000 less than the amount of the tender for work in accordance with the Engineer's method of construction. All of the other tender documents, namely the conditions of contract, specification, bill of quantities and drawings were, *mutatis mutandis,* incorporated in the alternative offer.

(5)   The alternative offer was, however, subject to the qualification that the material to be dredged would conform to the description given in the site investigation report, namely that it would comprise sand, silt and soft clay, and would exclude boulders.

(6)   At a meeting with the Engineer on 5 April 1997 the Contractor explained that the reason for the qualifications to his alternative offer was that the very substantial saving in cost associated with that offer could be achieved only by use of a suction dredger, namely the 'Gemini', which was ideally suited to the type of sea bed disclosed in the site investigation report, but which would be unable to work in hard clay, or to dredge boulders.

(7)   Following discussions at the said meeting it was agreed that within the alternative offer hard clay would be defined as material having a shear strength in excess of $0.2 \, \text{N}/\text{mm}^2$ and that boulders would be defined as being hard stones exceeding 250 mm in any direction.

(8)   An extra-over rate for dredging hard clay was agreed at £120 per cubic metre, but no provision was made for dealing with boulders, although dredging such material was expressly excluded from the contractor's obligations.

(9)   A formal contract, based upon the alternative offer and incorporating the agreements referred to in (7) and (8) above, was signed and sealed by both parties on 19 April 1997.

NARRATIVE

(10)   The outfall sewer extends from its westward end, chainage 0 metres, for a length of 1600 metres to the east. The first 100 metre

length was to be laid, under the alternative scheme adopted in the contract, in open trench excavation, part of which would be in tidal working. The remaining length from chainage 100 to chainage 1600 was to be laid in a dredged trench.

(11) On 26 April 1997 the Contractor, pursuant to clause 14 of the ICE Conditions, submitted to the Engineer his programme for the works, which provided for commencement of the works on 9 May 1997 and for completion on 5 August 1997.

(12) The Engineer consented to the Contractor's clause 14 programme by letter dated 2 May 1997.

(13) Work was commenced in accordance with the programme on 9 May 1997, by which date the 'Gemini' was in position at chainage 100 on the line of the outfall.

(14) On 20 May 1997, on which date dredging reached chainage 280, the 'Gemini' encountered conditions which prevented its operation. On examination of the sea bed by divers, it was found that boulders up to 400 mm in size had been uncovered at a depth of 0.8 metres below sea-bed level. Later investigation indicated that the stratum in which boulders were present extended to at least 3.5 metres below sea-bed level, which was the formation level for the sewer trench.

(15) By letter dated 21 May 1997, sent by fax on that day to the Engineer, the Contractor gave notice under clause 12(1) and submitted a claim under clause 12(2) of the conditions of contract, wherein the Contractor stated:

(1) That physical conditions and/or artificial obstructions had been encountered, which conditions and/or obstructions could not reasonably have been foreseen by an experienced contractor.

(2) That notice is hereby given pursuant to clause 12(2) and 52(4) of the conditions of contract of our intention to claim additional payment in respect of additional costs incurred by reason of the presence of boulders within the material to be dredged below the sea bed, and by reason of delays occasioned thereby. Pursuant to clause 44(1) it is intended to claim an extension of time in respect of the delay cause by the said conditions.

(3) That the effect of the said conditions and/or obstructions is to prevent the dredger 'Gemini' from operating efficiently or at all.

221

(4) That the Contractor proposes to commission the back-hoe dredger 'Colossus', currently berthed in Rotterdam, to proceed immediately to the site of the works and to dredge from chainage 280 until such time as sea bed conditions revert to those indicated in the site investigation report incorporated in the contract.

(5) That while it is not possible at that stage to foresee the full extent of the expected delay to the works, which delay is dependent upon the extent of the said conditions and/or obstructions, it will take one week from the date of your approval of our proposals for the 'Colossus' to arrive on site. Thereafter the Contractor expects the rate of working of the 'Colossus' to be similar to that of the dredger 'Gemini' notwithstanding the presence of boulders, because of the substantially greater power of the 'Colossus'.

(6) That the cost of commissioning the 'Colossus' including travel to and from the site is £360 000. Operating time on site will cost £1440 per hour and standing time £960 per hour.

(7) That the Contractor intends to maintain contemporaneous records of additional costs incurred by reason of the matters specified in item 4 above.

(8) That meanwhile the dredger 'Gemini' is being moved to chainage 1600 from which point it will, if sea bed conditions are suitable, dredge in a westerly direction towards the shore.

(9) That in order to minimise delay and additional costs your early approval of the above proposals is requested.

(16) In a covering letter to the formal notification referred to in paragraph (15) above the Contractor acknowledged that, because the possibility that boulders might be encountered had been discussed at the pre-contract meeting on 5 April 1997, it was arguable that such physical conditions had been foreseen, and hence that a claim under clause 12 was inadmissible. However, the contract expressly excluded the dredging of boulders and therefore the Contractor was entitled to be reimbursed for the additional costs occasioned thereby, either under clause 12 or as a variation under clause 51/52. Because clause 12 set out a procedure giving the Engineer full control over the method of dealing with the conditions/obstructions that had been encountered, the Contractor had submitted his claim under that clause.

(17) In the absence of a reply from the Engineer the Contractor telephoned on 24 May 1997 to emphasise the urgency of his decision

and to point out that, while the 'Colossus' was currently available, she might be commissioned at any time for other work, in which event further delay would result while a suitable alternative was located.

(18)   By letter dated 27 May 1997 sent by fax the Engineer approved the Contractor's proposals, as outlined in paragraph (15) above, as a means of dealing with the situation that had arisen, but rejected the claim for payment on the ground that the Contractor ought to have allowed in his rates for dealing with boulders.

(19)   In order to minimise delay the Contractor commissioned the 'Colossus' by fax immediately upon receipt of the Engineer's instruction (on 27 May 1997) and replied by letter on that day indicating the Contractor's intention to pursue the claim notwithstanding the Engineer's rejection thereof.

(20)   The dredger 'Colossus' arrived on site on 1 June 1997 and was engaged in dredging until 8 June 1997, by which date she had dredged from chainage 280 to chainage 420. At that point the sea bed was free of boulders and was of the sandy/silty nature described in the site investigation report.

(21)   Meanwhile the 'Gemini' had dredged from chainage 1600 to 1300, at which point hard clay was encountered at a depth 1 metre below the sea bed.

(22)   The Contractor accordingly instructed the 'Gemini' to move to chainage 420 and to dredge from there in an easterly direction, and instructed the 'Colossus' to move to chainage 1300 and to dredge in a westerly direction.

(23)   By 22 June 1997 the 'Colossus' had dredged from chainage 1300 to chainage 1000, at which point the stratum of hard clay disappeared and the sea bed reverted to the silts, sands and soft clays as described in the site investigation report. The 'Gemini' had dredged from chainage 420 to chainage 700 without further difficulty.

(24)   Accordingly the Contractor put the 'Colossus' on stand-by and instructed the 'Gemini' to continue dredging in an easterly direction from chainage 700 to chainage 1000, which work was completed by 10 July 1997 without encountering any further obstruction.

(25)   Meanwhile excavation in the shore and tidal area had been completed, as had pipe fabrication in preparation for launching. The pipe was laid in the prepared trench on 15/16 July 1997. Notwith-

standing the obstructions referred to herein, work was completed by the contract completion date of 5 August 1991.

(26)   Laboratory tests on representative samples of the hard clay excavated by the 'Colossus' between chainages 1300 and 1000 showed its shear strength to be in the range of 0.3 to 0.4 N/mm^2 and therefore to be 'hard clay' within the definition of that term as prescribed in the contract.

THE CLAIMS

(1)   The Contractor claims the following:

(1) Pursuant to clause 12(6) of the contract, the costs incurred in commissioning the dredger 'Colossus' and in hiring that dredger for the period from 12.00 noon on 1 June 1997 until 16.00 on 8 June 1997, plus 10% margin on the nett cost to cover off-site overheads and profit.

(2) The costs of additional hire of the dredger 'Gemini' occasioned by moving from chainage 280 to 1600 and from chainage 1300 to 420, together with 10% margin as above.

(3) Pursuant to the terms of the contract, an extra-over payment in respect of dredging hard clay falling within the contractual definition of that term.

(4) Pursuant to clause 60(7), interest on the above claims, full details of which were submitted with monthly valuation number 3 on 3 August 1997. Payment became due 28 days later, i.e. by 31 August 1997. Interest on the total claim, namely £1 092 600 at 2% above base rate, compounded monthly, is claimed for the period from 31 August 1997 until 31 December 1998.

(2)   Particulars of the above claims are given below:

Cost of commissioning the 'Colossus'	360 000
Hire of ditto: 172 hours at £1440/hour	247 680
Plus off-site overheads and profit at 10%	60 768
	£668 448

Additional hire of the 'Gemini':

Two moves at 8 hours per move: 16 hours × £720 per hour	11 520
Off-site overheads and profit at 10%	1 152
	£12 672

Excavation of hard clay:
Chainage 1300–1000: trench 5 m wide × 3.5 m deep
5,250 cu m at £120 per m   630 000

Total   £1 311 120

Interest* calculated at 2% above base rate, compounded
monthly, for the period from 31 August 1997 to
31 December 1998:

$$1\,311\,120 \times (2.4959891/2.2095642 - 1) = £169\,960$$

Total claim   **£1 481 080**

*Interest calculation is based on factors tabulated in *Arbitration Practice in Construction Contracts* (4th edn) by Douglas Stephenson (Blackwell Science).

CHAPTER NINE
THE ICE DESIGN AND CONSTRUCT CONTRACT

9.1 Introduction

First published in October 1992, the ICE Design and Construct Contract was updated in March 1995 by the publication of a *Guidance Note* in which a new clause, numbered 71, was inserted in order to take account of the *Construction (Design and Management) Regulations 1994*. The form of contract represents a major departure from the long-established principle of all previous ICE conditions and their derivatives: namely that the engineer performs his functions under a completely separate contract with the employer from that of the contractor.

The merits of the traditional separation of design functions from those of construction have, however, been questioned during the 1990s and to a lesser extent during earlier decades. Does it really benefit the employer to have an engineer and a contractor who, instead of collaborating in a project, may be isolated from, and in some cases antagonistic towards, one another? In other fields of engineering, such as mechanical, electrical, aeronautical and automobile, design and construction are usually performed by a single organisation with the aim to achieve overall efficiency in both aspects of a given project and thereby to prosper in a competitive market.

9.1.1 Development of the form

Some of the earlier excursions from traditional methods resulted from a willingness of employers to consider, in addition to tenders for the engineer's own design, alternatives offered by contractors. Contracts awarded on that basis included, in the late 1950s, Gladesville Bridge in Sydney, in which a precast concrete segmental arch was offered by a tenderer as an alternative to the engineer's steel cantilever design. Although successful in that instance and in many others, the approach is clumsy in that the 'official' design must be developed in sufficient detail to provide a basis for tendering; and the time and effort put into that work is wasted if an alternative design proves to be more competitive.

Recognition of that disadvantage has no doubt been a factor in the growing popularity of the 'package deal' in which tenders are invited for achieving specified performance criteria. The advent of privately financed road, bridge, and similar public projects in an area that was at one time the sole domain of the public authority has promoted that growth.

The ICE Design and Construct Conditions of Contract represent a first attempt by the Conditions of Contract Standing Joint Committee of the three sponsoring bodies – the ICE, the FCEC (since superseded by the CECA) and the ACE – to formalise and standardise a contract based on the integration of design and construction. Although it was introduced as a completely new form, the publishers' description refers to it as having been developed from the Sixth Edition, as clearly it has. Its format remains unchanged wherever possible, as indeed has the clause numbering system. Where clauses of the Sixth Edition are unnecessary the clause number is noted in the Design and Construct Contract as being 'not used' in order to maintain the numbering system that has remained substantially unchanged since the first edition of the ICE Conditions was published in 1945.

9.1.2 Fundamental changes

Two fundamental changes in the Design and Construct Contract, which have repercussions throughout the form, result from the altered function of 'the Engineer', who becomes a member of the contractor's team; and the likely omission of a bill of quantities. The first of these changes is accommodated by the introduction of an 'Employer's Representative', who exercises some, but not all, of the powers of 'the Engineer' in a conventional contract.

It is to be presumed that generally a design and construct contract is awarded to the tenderer who offers the most favourable price for meeting a performance specification – possibly for example a specified area, equipment and environmental standards in the case of a building, or specified geometric design and load-bearing

standards in the case of a road, bridge or tunnel connecting specified points. In either such case, and in many other types of project,
the need for a bill of quantities may not arise. The main functions of
such a document in conventional contracts are to provide a basis:

- for tendering and for evaluation of tenders;
- for monthly valuations and for evaluating the final account; and
- for evaluating variations.

In the case of a design and construct contract the first requirement
disappears – the quantities of work required to achieve the specified
requirements being irrelevant and of no interest to the employer.
Furthermore those quantities are dependent upon the detailed
design to be prepared by the contractor when appointed. The
second function, namely monthly valuations, must necessarily be
achieved by some other means – for example, by defining various
stages of completion in the building or sections of the work in a
road, bridge or tunnel contract (earthworks, drainage, sub-base
etc.). The third function, evaluation of variations, is necessarily
more conjectural than is the case where a bill of quantities is
available. However, against that disadvantage must be set the
advantage that the need for variations should in most cases be
substantially reduced – those needed to cover design errors,
omissions, delays, and test requirements, which the engineer in a
conventional contract would order, become an internal matter
within the contractor's organisation and of no concern to the
employer.

It follows that in many instances the opportunities for claims
under the Design and Construct Contract will be fewer than under
the Sixth Edition. Of the 22 'claims' clauses in the Sixth Edition no
fewer than eight (7(4), 13(3), 14(8), 17(2), 52(4), 55(2), 56(2) and 59(4))
have no counterpart in the Design and Construct Contract. Also the
opportunities for claiming in the remaining 'claims' clauses are
greatly reduced in comparison with the Sixth Edition.

Against the many advantages to be gained from integrating
design with construction must be set at least one major disadvantage. The design and construct procedure necessarily
requires that each tenderer must prepare a preliminary design for
the project in sufficient detail to enable construction work to be
priced. This is a considerably more costly procedure than merely
pricing the work involved in constructing to an engineer's design,
especially where that design includes a detailed bill of quantities.

The disadvantage is exacerbated in cases where the site investigation requirements differ from one possible solution to another – for example where a river crossing may be in the form either of a suspension bridge or of a viaduct, or possibly even a tunnel.

In order to limit the amount of unproductive work to be performed by unsuccessful tenderers it is usually necessary to restrict the number of schemes to be prepared, either at the stage of preparing a selected list of tenderers or following upon receipt of outline proposals.

9.2   Claims and the employer's representative

Clause 1(1)(c) of the Design and Construct Contract states:

> 'Employer's Representative' means the person appointed by the Employer to act as such for the purposes of the Contract or any other person so appointed from time to time by the Employer and notified in writing as such to the Contractor.

while clause 2 defines the duties and authority of the employer's representative, his identity, his powers of delegation, and the format of his instructions, as follows:

(1)   (a)   The Employer's Representative shall carry out the duties and may exercise the authority specified in or necessarily to be implied from the Contract.
(b) Except as expressly stated in the Contract the Employer's Representative shall have no authority to amend the Contract nor to relieve the Contractor of any of his obligations under the Contract.
(2)   Within 7 days of the award of the Contract and in any event before the Commencement Date the Employer shall notify to the Contractor in writing the name of the Employer's Representative. The Employer shall similarly notify the Contractor of any replacement of the Employer's Representative.
(3)   The Employer's Representative may from time to time delegate to any person (including assistants appointed under sub-clause (4) of this Clause) any of the duties and authorities vested in him and he may at any time revoke such delegation.
Any such delegation
(a)   shall be in writing and shall not take effect until such time as a copy thereof has been delivered to the Contractor or a representative appointed under Clause 15

(b) shall continue in force until such time as the Employer's Representative shall notify the Contractor or his representative in writing that the same has been revoked and

(c) shall not be given in respect of any decision to be taken or certificate to be issued under Clauses 12(6) 15(2)(b) 44 46(3) 48 60(4) 61 or 65.

(4) (a) The Employer's Representative may appoint any number of assistants for the carrying out of his duties under the Contract. He shall notify to the Contractor the names duties and scope of authority of such assistants.

(b) Assistants may be appointed specifically to watch the construction of the Works.

(c) Assistants shall have no authority to relieve the Contractor of any of his duties or obligations under the Contract nor except as provided under sub-clause (3) of this Clause

(i) to order any work involving delay or any extra payment by the Employer or

(ii) to make any variation of or in the Works.

(5) (a) Instructions given by the Employer's Representative or by any person exercising delegated duties and authorities under sub-clause (3) of this Clause shall be in writing. Provided that if for any reason it is considered necessary to give any such instruction orally the Contractor shall comply therewith.

(b) Any such oral instruction shall be confirmed in writing by or on behalf of the Employer's Representative as soon as is possible in the circumstances. Provided that if the Contractor confirms in writing any such oral instruction which is not contradicted in writing by or on behalf of the Employer's Representative forthwith it shall be deemed to be an instruction in writing by the Employer's Representative.

(c) Upon the written request of the Contractor the Employer's Representative or the person exercising delegated duties or authorities under sub-clause (3) of this Clause shall specify in writing under which of his duties and authorities any instruction is given.

(6) If the Contractor is dissatisfied by reason of any act instruction or decision of any assistant appointed under this Clause he shall be entitled to refer the matter to the Employer's Representative for his decision.

Separation of the engineer's two functions in a conventional contract, namely those of design and of supervision of construction, raises a question whether or not the employer's representative in a design and construct contract has any duty towards the contractor. His title implies that, as the employer's representative, his sole concern is with the interests of the employer, for whom he is undoubtedly an agent. Impliedly he no doubt has a duty to be honest and possibly to be fair; although whether or not that duty would extend, for example, to pointing out to the contractor opportunities that may arise for claiming additional payment is open to question.

In contrast to the engineer in the Sixth, but not in earlier editions, the employer's representative is not necessarily a chartered engineer. It follows that the contractor in a design and construct contract cannot rely on the employer's representative complying with the ethical standards required of members of the ICE. It is however to be hoped and indeed expected that the words of Lord Morris of Borth-y-Gest in *Sutcliffe* v. *Thackrah* (1974):

> 'Being employed by and paid by the owner [the Engineer] unquestionably has in divers ways to look after the interests of the owner. In doing so he must be fair and he must be honest. He is not employed by the owner to be unfair to the contractor.'

will be as applicable to an employer's representative as to an architect or an engineer – on the basis that he is similarly in a position of authority requiring the exercise of a quasi-judicial power.

The powers of the employer's representative to delegate his functions correspond closely to those of an engineer in a conventional contract, as do the contractor's rights to refer an assistant's decision with which the contractor is dissatisfied to the employer's representative himself. Similarly the limitation in a conventional contract of the engineer's powers to delegate certain functions to a representative or an assistant (namely those under clauses 12(6), 44, 46(3), 48, 60(4), 61 and 63) reappear as limitations imposed on the employer's representative in clauses similarly numbered in the Design and Construct Contract, except that clause 63 of the Sixth Edition has its counterpart in clause 65 of the Design and Construct Contract.

A welcome change is introduced in clause 66 (settlement of disputes) in that the provision in earlier forms for reference of disputes to the engineer for his formal decision under the clause no longer applies. Clearly the replacement of the engineer in a conventional contract by an employer's representative in the design and construct contract was a factor, but the solution to the problem is one that has been urged for many decades by those who believe the two-stage procedure of the conventional clause 66 to be otiose. Instead of the dispute being referred to the engineer – whose decision in all probability gave rise to the dispute – it is now referred directly to conciliation. Hence a decision is now to be sought from an engineer who is unquestionably impartial.

9.3 *Contractual claims*

9.3.1 Cost

The definition of 'cost' in clause 1(5) of the Design and Construct Contract is:

> The word 'cost' when used in the Conditions of Contract means all expenditure including design costs properly incurred or to be incurred whether on or off the Site and overhead finance and other charges (including loss of interest) properly allocatable thereto but does not include any allowance for profit.

Except for the necessary inclusion of design costs, and the addition of loss of interest, this definition is identical with that in the Sixth Edition. Some of the clauses in that edition in which the word is used do not appear in the Design and Construct Contract, in particular, clauses 7, 13, 14 and 17 do not appear as 'claims' clauses.

9.3.2 Contract documents

Clause 5 of the Design and Construct Contract provides as follows:

(1) (a) The several documents forming the Contract are to be taken as mutually explanatory of one another.
(b) If in the light of the several documents forming the Contract there remain ambiguities or discrepancies between the Employer's Requirements and the Contractor's Submission the Employer's Requirements shall prevail.
(c) (i) Any ambiguities or discrepancies within the Employer's Requirements shall be explained and adjusted by the Employer's Representative who shall thereupon issue to the Contractor appropriate instructions in writing.
(ii) Should such instructions involve the Contractor in delay or disrupt his arrangements or methods of construction so as to cause him to incur cost beyond that reasonably to have been

foreseen by an experienced contractor at the time of the award of the Contract then the Employer's Representative shall take such delay into account in determining any extension of time to which the Contractor is entitled under Clause 44 and the Contractor shall subject to Clause 53 be paid in accordance with Clause 60 the amount of such cost as may be reasonable. Profit shall be added thereto in respect of the design and construction of any additional permanent or temporary work.

(d)   Any ambiguities or discrepancies within the Contractor's Submission shall be resolved at the Contractor's expense.

(2)   Upon the award of the Contract the Employer shall assemble two complete copies of the Contract of which one copy shall be supplied to the Contractor free of charge.

It is to be expected in a design and construct contract that the majority of the drawings will be prepared by the contractor, who will of course be responsible for rectifying his own errors and for the financial consequences of such errors. Only those drawings needed to illustrate the employer's requirements – site plans, etc. – are likely to be supplied by the employer and hence to be a potential basis of claim under this clause.

9.3.3   Late issue of further information

Clause 6 of the Design and Construct Contract provides as follows:

(1)   (a)   The Contractor shall give the Employer's Representative notice in writing of any further information from the Employer that is required for the design and/or construction of the Works and to which the Contractor is entitled under the Contract.

(b)   Should the Employer's Representative fail to issue any such information within a reasonable period following notice he shall take that failure into account in determining any extension of time to which the Contractor is entitled under Clause 44 and the Contractor shall subject to Clause 53 be paid in accordance with Clause 60 any reasonable cost which may arise from the failure.

(2)   (a)   The Contractor shall except as may otherwise be provided in the Contract submit to the Employer's Representative such designs and drawings as are necessary to show the general arrangement of the Works and that the Works will comply with the Employer's Requirements.

Construction to such designs and drawings shall not commence until the Employer's Representative has consented thereto.

(b)   If in the opinion of the Employer's Representative any such design or drawing does not comply with the Employer's Requirements or with any other provision of the Contract he shall so inform the Contractor in writing giving his reasons and may withhold his consent thereto until the Contractor has re-submitted the design or drawing with appropriate modifications.

(c)   The Contractor shall notify the Employer's Representative if he later wishes to modify any design or drawing to which consent has already been given and shall submit the modified design or drawing for further consent.

(d)   Should the Employer's Representative fail within a reasonable period following the submission or re-submission of any design or drawing under this Clause to notify the Contractor either that he consents thereto or that he is withholding his consent he shall take such failure into account in determining any extension of time to which the Contractor is entitled under Clause 44 and the Contractor shall subject to Clause 53 be paid in accordance with Clause 60 any reasonable cost which may arise from such failure.

(3)   The Contractor shall supply to the Employer's Representative two copies of such other designs drawings specifications documents and information as the Employer's Representative may require.

Since the majority of the drawings required for a design and construct contract are likely to be the responsibility of the contractor, the likelihood that claims will arise from late issue of drawings is correspondingly reduced. Another type of claim is however introduced: namely that resulting from the employer's representative causing delay in consenting to the contractor's submissions. In such cases the contractor is entitled to 'reasonable cost' and an extension of time.

9.3.4   Unforeseeable conditions and obstructions

Clauses 11 and 12 of the Design and Construct Contract do not differ materially from the corresponding clauses in the Sixth Edition. (See Chapter 3, section 3.5 above.)

9.3.5   Satisfaction of the engineer

The Design and Construct Contract contains no clause corresponding to clause 13 of the Sixth and earlier editions, presumably because it would imply an inappropriate degree of control over the contractor's method of working.

9.3.6 Construction programme

Except for a minor degree of re-phrasing to a more concise format, clause 14 of the Design and Construct Contract does not differ substantially from its counterpart in the Sixth Edition in the mechanics of submission and approval of the construction programme. There is, however, no express provision for claims in respect of delay by the employer's representative in giving consent.

9.3.7 Setting-out errors

Clause 17 of the Sixth Edition is omitted from the Design and Construct Contract, presumably on the basis that all, or nearly all, of the setting-out will be the responsibility of the contractor. While it is not difficult to envisage a situation in which the employer's representative in a design and construct contract provides incorrect data to the contractor – for example, the precise location of a factory in a green field site – any such error could be dealt with by the issue of a variation order, in respect of which the contractor would be entitled to payment and to an extension of time in the usual way.

9.3.8 Boreholes and exploratory excavation

Clause 18 of the Design and Construct Contract states:

(1)   If during the performance of the Works the Contractor considers it necessary or desirable to make boreholes or to carry out exploratory excavations or investigations of the ground he shall apply to the Employer's Representative for permission so to do giving his reasons and details of his proposed methods. Such permission shall not unreasonably be withheld.
The Contractor shall comply with any conditions imposed by the Employer's Representative in relation thereto and shall furnish the Employer's Representative with copies of all information records and test results arising therefrom and of any expert opinion as may be provided in connection therewith.

(2) The cost of making such boreholes and carrying out such investigations and of all other matters connected therewith including making good thereafter to the satisfaction of the Employer's Representative shall be borne by the Contractor. Provided that if in the opinion of the Employer's Representative the boreholes excavations and investigations are a necessary consequence of

(a) a situation arising under Clause 12 they shall be paid for in accordance with that Clause or

(b) a variation ordered under Clause 51 they shall for the purposes of payment be treated as part of that variation and priced in accordance with Clause 52.

(3) If during the performance of the Works the Employer's Representative requires the Contractor to make boreholes or to carry out exploratory excavations or investigations of the ground such requirements shall be ordered in writing and shall be deemed to be a variation ordered under Clause 51 unless a Contingency or Prime Cost Item in respect of such anticipated work is included in the Contract.

The only two situations in which there may be a claim for the cost of site investigations are, logically, where unforeseeable ground conditions are encountered giving rise to a claim under clause 12, and where the employer's representative orders a variation.

9.3.9 Damage occasioned by excepted risks

Clause 20 of the Design and Construct Contract states:

(1) (a) The Contractor shall save as in paragraph (b) hereof and subject to sub-clause (2) of this Clause take full responsibility for the care of the Works and for materials plant and equipment for incorporation therein from the Commencement Date until the date of issue of a Certificate of Substantial Completion for the whole of the Works when the responsibility for the said care shall pass to the Employer.

(b) If the Employer's Representative issues a Certificate of Substantial Completion for any Section or part of the Permanent Works the Contractor shall cease to be responsible for the care of that Section or part from the date of issue of that Certificate of Substantial Completion when the responsibility for the care of that Section or part shall pass to the Employer. Provided always that the Contractor shall remain responsible for any damage to such completed work caused by or as a result of his other activities on the Site.

(c) The Contractor shall take full responsibility for the care of any outstanding work and materials plant and equipment for incorporation therein which he undertakes to finish during the Defects Correction Period until such outstanding work has been completed.

(2) Risks for which the Contractor is not liable are loss and damage to the extent that they are due to

(a) the use or occupation by the Employer his agents servants or other contractors (not being employed by the Contractor) or any part of the Permanent Works

(b) any fault defect error or omission in the design of the Works for which the Contractor is not responsible under the Contract

(c) riot war invasion act of foreign enemies or hostilities (whether war be declared or not)

(d) civil war rebellion revolution insurrection or military or usurped power

(e) ionizing radiations or contamination by radioactivity from any nuclear fuel or from any nuclear waste from the combustion of nuclear fuel radioactive toxic explosive or other hazardous properties of any explosive nuclear assembly or nuclear component thereof and pressure waves caused by aircraft or other aerial devices travelling at sonic or supersonic speeds.

(3) (a) In the event of any loss or damage to

(i) the Works or any Section or part thereof or

(ii) materials plant or equipment for incorporation therein

while the Contractor is responsible for the care thereof (except as provided in sub-clause (2) of this Clause) the Contractor shall at his own cost rectify such loss or damage so that the Permanent Works conform in every respect with the provisions of the Contract. The Contractor shall also be liable for any loss or damage to the Works occasioned by him in the course of any operations carried out by him for the purpose of complying with his obligations under Clauses 49 and 50.

(b) Should any loss or damage arise from any of the Excepted Risks defined in sub-clause (2) of this Clause the Contractor shall if and to the extent required by the Employer's Representative rectify the loss or damage at the expense of the Employer.

(c) In the event of loss or damage arising from an Excepted Risk and a risk for which the Contractor is responsible under sub-clause (1)(a) of this Clause then the cost of rectification shall be apportioned accordingly.

As might be expected, the contractor under a design and construct contract must accept risks arising from that part of the design for which he is responsible. Apart from that change, the provisions of clause 20 of the Design and Construct Contract do not differ substantially from those of the Sixth Edition.

9.3.10 New Roads and Street Works Act 1991

Many of the claims situations envisaged by clause 27 of the Sixth and earlier editions arise from variations, some of which may be needed to correct design errors or inadequacies. Such variations are the responsibility of the contractor under a design and construct contract, and only those originating from a change ordered by the employer form the subject of claims. The omission from the Design and Construct Contract of any special provision for the operation of the 1991 Act appears therefore to be logical – any claim arising from the effect of the Act on a variation ordered by the employer being recoverable under clause 51.

9.3.11 Provision of facilities for other contractors, disposal of fossils etc.

Clauses 31 and 32 of the Design and Construct Contract do not differ materially from the provisions of the same clauses in the Sixth Edition.

9.3.12 Cost of samples and of tests

Clause 36 of the Design and Construct Contract states:

(1) The Works shall be designed constructed and completed in accordance with the Contract and where not expressly provided otherwise in the Contract in accordance with appropriate standards and standard codes of practice.

(2) All materials and workmanship shall be of the respective kinds described in the Contract or where not so described shall be appropriate in all the circumstances.

(3) (a) Further to his obligations under Clause 8(3) the Contractor shall submit to the Employer's Representative for his approval proposals for checking the design and setting out of the Works and testing the materials and workmanship to ensure that the Contractor's obligations under the Contract are met.

(b) The Contractor shall carry out the checks and tests approved under sub-clause (3)(a) of this Clause or elsewhere in the Contract and such further tests as the Employer's Representative may reasonably require.

(4)   The Contractor shall provide such assistance and such instruments machines labour and materials as are normally required for examining measuring and testing any work and the quality weight or quantity of any materials used and shall supply samples of materials before incorporation in the Works for testing as may be required by the Employer's Representative.

(5)   All samples shall be supplied by the Contractor at his own cost if the supply thereof is clearly intended by or provided for in the Contract but if not then at the cost of the Employer.

(6)   Whenever a variation is ordered or consented to by the Employer's Representative the Contractor shall consider whether any tests would be affected by or be appropriate in relation thereto and shall so inform the Employer's Representative without delay. Any proposal for amended or additional tests shall be submitted as soon as possible.

(7)   Unless the Contract otherwise provides the cost of making any test shall be borne by the Contractor if such test is

   (a)   proposed by the Contractor under Clause 8(3) or sub-clause (3)(a) of this Clause or

   (b)   clearly intended by or provided for in the Contract.

If any test is carried out pursuant to sub-clause (3)(b) of this Clause the cost of such test shall be borne by the Contractor if the test shows the workmanship or materials not to be in accordance with the provisions of the Contract but otherwise by the Employer.

Responsibility for deciding upon the sampling and testing needed to ensure that materials and workmanship are satisfactory rests, in the Design and Construct Contract, on the contractor, who must submit his proposals to, and obtain the approval of, the employer's representative. Thereafter he must implement the approved proposals, all at his own expense.

However, the employer is also entitled to require additional sampling, the cost of which is dealt with in the same way as that under the Sixth Edition – namely, the contractor must at his own cost supply samples 'clearly intended by or provided for in the contract', other samples entitling the contractor to payment of cost.

The cost of testing is dealt with in the same way, with the proviso that the cost of any test that shows the workmanship or materials not to be in accordance with the provisions of the contract is borne by the contractor.

Another situation arises where additional testing becomes necessary as a result of variation ordered by the employer's representative. Samples for such tests and the cost of testing would clearly entitle the contractor to additional payment since they could not have been provided for in the contract and it appears from the wording of the clause that in such circumstances the contractor is entitled to payment even if the test shows workmanship or materials to be unsatisfactory.

9.3.13 Cost of uncovering work

Clause 38 of the Design and Construct Contract contains similar provisions to those of the Sixth Edition, enabling the employer's representative to examine work before it is covered up and where necessary require the contractor to uncover work if an opportunity to inspect has not been provided.

9.3.14 Suspension of work

Clause 40 of the Design and Construct Contract makes provisions similar to those of clause 40 of the Sixth Edition for a suspension of work to be ordered by the employer's representative, and provides similar safeguards to the contractor.

9.3.15 Possession of the site

Similarly, clause 42 of the Design and Construct Contract does not differ materially from its counterpart in clause 42 of the Sixth Edition.

9.3.16 Rate of progress

Here again, clause 46 of the Design and Construct Contract reflects the provisions of the same clause of the Sixth Edition.

9.3.17 Outstanding work and defects

The Design and Construct Contract follows the terminology used in the Sixth, but not in earlier editions, of 'defects correction' rather than 'maintenance' as describing the period following substantial completion. The contractor's responsibilities during that period are similar to those under the Sixth Edition.

9.3.18 Contractor to search

Clause 50 of the Design and Construct Contract does not differ materially from the same clause of the Sixth Edition.

9.3.19 Use of contingency and prime cost items

Clause 58 of the Design and Construct Contract states:

(1) The Contractor shall not commence work on any Contingency or Prime Cost Item until he has secured the consent thereto of the Employer's Representative which consent shall not unreasonably be withheld.
(2) Contingencies and Prime Cost Items shall be valued and paid for in accordance with Clause 52 or as the Contract otherwise provides. The percentage to be used for overheads and profit in adjusting the Prime Cost element of any such item shall be the figure stated therefor in the Appendix to the Form of Tender.

Clause 59, which in the Sixth and earlier editions provides for nominated sub-contracts, is 'not used' in the Design and Construct Contract. The term 'contingency' has been introduced in the Design and Construct Contract to replace 'provisional sum' as used in the Sixth and in earlier editions. Its definition, in clause 1(1)(j), is almost identical to the definition of its counterpart in clause 1(1)(l) of the Sixth Edition.

As in the Sixth Edition, the definitions make it clear that a prime cost sum will be used, while a contingency may be used. In both cases the employer's representative retains control over their use.

The principal, and very welcome, change in the Design and Construct Contract is the elimination of provision for nominated sub-contracts – presumably because the circumstances in which such provision might be needed are unlikely to arise in such a contract.

9.3.20 Interest on overdue payments

Apart from the inevitable substitution of 'employer's representative' for 'engineer', clause 60 (7) of the Design and Construct Contract makes identical provision to that of the Sixth Edition for the payment of interest, compounded monthly at 2% over base rate, on sums which the employer's representative fails to certify or the employer fails to pay.

9.4 Variations

9.4.1 Power to vary

Clause 51 of the Design and Construct Contract states:

(1) The Employer's Representative shall have power after consultation with the Contractor's Representative to vary the Employer's Requirements. Such variations may include additions and/or omissions and may be ordered at any time up to the end of the Defects Correction Period for the whole of the Works.

(2) All variations shall be ordered in writing but the provisions of Clause 2(5) in respect of oral instructions shall apply.

(3) No variation ordered under this Clause shall in any way vitiate or invalidate the Contract but the fair and reasonable value (if any) of all such variations shall be taken into account in ascertaining the amount of the Contract Price except to the extent that such variation is necessitated by the Contractor's default.

Although the employer retains the power, through his representative, to vary his requirements, he may do so only after consultation with the contractor's representative.

An important category of variations in a conventional contract, namely those needed to correct design errors, discrepancies and inadequacies, is eliminated in a design and construct contract. This is because such matters are dealt with internally within the contractor's organisation and are of no concern to the employer. Hence the absence, in clause 51 of the Design and Construct Contract, of a *duty* to vary. As in the Sixth Edition, variations do not vitiate or invalidate the contract but their value is taken into account in the contract price.

9.4.2 Valuation of ordered variations

Clause 52 of the Design and Construct Contract states:

(1)   When requested by the Employer's Representative the Contractor shall submit his quotation for the work as varied and his estimate of any delay. Wherever possible the value and delay consequences (if any) of each variation shall be agreed before the order is issued or before work starts.

(2)   In all other cases the valuation of variations ordered by the Employer's Representative in accordance with Clause 51 shall be ascertained as follows.

(a)   As soon as possible after receipt of the variation order the Contractor shall submit to the Employer's Representative

(i)   his quotation for any extra or substituted works necessitated by the variation having due regard to any rates or prices included in the Contract and

(ii)   his estimate of any delay occasioned thereby and

(iii)   his estimate of the cost of any such delay.

(b)   Within 14 days of receiving the said submissions the Employer's Representative shall

(i)   accept the same or

(ii)   negotiate with the Contractor thereon.

(c)   Upon reaching agreement with the Contractor the Contract Price shall be amended accordingly.

(d)   In the absence of agreement the Employer's Representative shall notify the Contractor of what in his opinion is a fair and reasonable valuation and thereafter shall make such interim valuations for the purposes of Clause 60 as may be appropriate.

(3)   The Employer's Representative may if in his opinion it is necessary or desirable order in writing that any additional or substituted work shall be executed on a daywork basis in accordance with the provisions of Clause 56(3).

One of the problems likely to arise from the use of the Design and Construct Contract is that of evaluating variations in the probable absence of a bill of quantities. The contract makes the sensible provision for agreement between the employer's representative and the contractor's representative of a price for the varied work, where possible covering also the consequences of any delay that may result from the variation, and for doing so as soon as possible after the variation order has been issued.

It should, however, be recognised by both parties that the contractor in such circumstances is not in competition with others for the varied work, and that he may be tempted to evaluate the work at a higher figure than if he were. Where the employer's representative is unable to agree what is in his opinion a reasonable price he has two alternatives, either to notify the contractor of that evaluation and to certify accordingly, or to order that the work be executed as daywork. In the former case the valuation would be open to review by an arbitrator under clause 66(7), while in the latter it would be determined under clause 56(3) of the contract.

9.4.3 Daywork

Clause 56(3) of the Design and Construct Contract states:

(3)   Where any work is carried out on a daywork basis the Contractor shall be paid for such work under the conditions and at the rates and prices set out in the daywork schedule included in the Contract or failing the inclusion of a daywork schedule he shall be paid at the rates and prices and under the conditions contained in the 'Schedule of Dayworks carried out incidental to Contract Work' issued by the Federation of Civil Engineering Contractors current at the date of the execution of the daywork.

The Contractor shall furnish to the Employer's Representative such records receipts and other documentation as may be necessary to prove amounts paid and/or costs incurred. Such returns shall be in the form and delivered at the times the Employer's Representative shall direct and shall be agreed within a reasonable time.

Before ordering materials the Contractor shall if so required submit to the Employer's Representative quotations for the same for his approval.

The use of daywork as a means of evaluating varied work is likely to become more prevalent in the absence of a bill of quantities, and no doubt many employer's representatives will expect to find a schedule of dayworks included in the contract documents as a provision for such evaluation. The provisions for evaluating work in this way do not differ materially from those in the Sixth and earlier editions. The FCEC *Schedule of Dayworks* has since publication of the Design and Construct form been replaced by a similar schedule published by the CECA.

9.4.4 Additional payments

Clause 53 of the Design and Construct Contract states:

(1)   If the Contractor intends to claim any additional payment pursuant to any Clause of these Conditions other than Clause 52(1) he shall give

notice in writing of his intention to the Employer's Representative as soon as may be reasonable and in any event within 28 days after the happening of the events giving rise to the claim.

Upon the happening of such events the Contractor shall keep such contemporary records as may reasonably be necessary to support any claim he may subsequently wish to make.

(2) Without necessarily admitting the Employer's liability the Employer's Representative may upon receipt of a notice under this Clause instruct the Contractor to keep such contemporary records or further contemporary records as the case may be as are reasonable and may be material to the claim of which notice has been given and the Contractor shall keep such records.

The Contractor shall permit the Employer's Representative to inspect all records kept pursuant to this Clause and shall supply him with copies thereof as and when the Employer's Representative shall so instruct.

(3) After the giving of a notice to the Employer's Representative under this Clause the Contractor shall as soon as is reasonable in all the circumstances send to the Employer's Representative a first interim account giving full and detailed particulars of the amount claimed to that date and of the grounds upon which the claim is based.

Thereafter at such intervals as the Employer's Representative may reasonably require the Contractor shall send to the Employer's Representative further up to date accounts giving the accumulated total of the claim and any further grounds upon which it is based.

(4) If the Contractor fails to comply with any of the provisions of this Clause in respect of any claim which he shall seek to make then the Contractor shall be entitled to payment in respect thereof only to the extent that the Employer's Representative has not been prevented from or substantially prejudiced by such failure in investigating the said claim.

(5) The Contractor shall be entitled to have included in any interim payment certified by the Employer's Representative pursuant to Clause 60 such amount in respect of any claim as the Employer's Representative may consider due to the Contractor provided that the Contractor shall have supplied sufficient particulars to enable the Employer's Representative to determine the amount due.

If such particulars are insufficient to substantiate the whole of the claim the Contractor shall be entitled to payment in respect of such part of the claim as the particulars may substantiate to the satisfaction of the Employer's Representative.

The opportunity has been taken in the Design and Construct Contract to remedy one of the well-known drafting defects in the Sixth

Edition and its predecessors, namely the inclusion of a clause relating to notice of claims under 'any Clause of these Conditions' as a sub-clause (52(4)) of a clause relating solely to variations. It now appears as a clause in its own right, namely as clause 53. This is made possible without any major disruption of the clause-numbering policy by including all matters relating to the ownership and vesting of the contractor's materials and equipment in a single clause, clause 54, leaving 53 available as the clause relating to claims notices in general.

The content of the new clause 53 is, however, almost identical to clause 52(4) of the Sixth Edition.

9.5 Delays and extension of time

The provisions of clause 47 of the Design and Construct Contract reflect those of the same clause in the Sixth and earlier editions, the only changes being those made necessary by the substitution of 'Employer's Representative' for 'Engineer'.

CHAPTER TEN
THE NEW ENGINEERING CONTRACT

10.1 Introduction

The 1980s saw an explosion of published forms of contract for use in the construction industries. The decade opened with the publication of a new edition of the JCT Standard Form of Building Contract. Many employers preferred to use the old 1963 edition and both versions were in regular use for some time. The JCT Intermediate Form appeared in 1984 to bridge the gap between the complexities of the main form and the inadequacies of the JCT Minor Works form.

Meanwhile the building industry was seeking the holy grail of a contract which would avoid confrontation and claims. In 1981 the JCT published its form of design and build contract. Its exponents passionately believed that it would provide the answer. They were wrong. Design/build contracting has become a popular and successful means of procuring a building, but it has not avoided claims.

In 1987 the JCT produced the Management Contract, another innovative form of contract which it was hoped would enable fast track construction to proceed in an atmosphere of co-operation and without the claims culture which seemed to pervade all large projects. Once again there are many examples of successful projects based on this form, but it cannot be said that it made claims redundant.

The civil engineering world did not produce new contracts as energetically as the builders, but a similar quest was being followed. Contracts were increasingly let on forms which were not originally designed for civil engineering projects, such as the forms published by the Institution of Chemical Engineers. The standard forms of Government contracts were also re-issued with new procedures designed to control claims and consequent cost overruns.

The Institution of Civil Engineers reacted to these developments by commissioning an entirely new approach to contracts. The original decision was made in 1985, and in 1988 work started on the detailed draft. The New Engineering Contract was published in a consultative edition in 1991 and it was used on a number of projects, particularly in South Africa. The First Edition was formally published in March 1993.

The contract provoked considerable controversy. It had been written in language which was quite different to any previous standard form, and opinion was divided as to whether this made the contract easier to read and understand, or more difficult. The contract was the subject of many unsatisfactory public debates between its proponents, who argued that it was a giant leap forward in concept, and those who argued that the detail contained drafting errors and ambiguities which were likely to lead to even more claims and litigation than the existing conventional contracts.

The New Engineering Contract ('NEC') did have one very influential supporter. Sir Michael Latham, in his report *Constructing the Team* published in 1994 not only praised the contract, but suggested that it should be adopted as the national standard contract for all construction projects, building as well as civil engineering.

If the contract was to have such wide use, it was no longer appropriate to have a name that was so wedded to the civil engineering industry. When the second edition was published in 1995 therefore its name was changed to the Engineering and Construction Contract, and it became known by its new initials 'ECC'. However it is still widely referred to by its old name and abbreviation 'NEC', and this book retains the more familiar terminology.

A further major amendment was published in April 1998 in an attempt to deal with the requirements of Part II of the Housing Grants, Construction and Regeneration Act 1996. This is discussed below at section 10.3.7.

10.2   The NEC Approach

The New Engineering Contract is suitable for use either in a conventional project, where the design is produced by the employer's design team, or in a project where the contractor is responsible for some or all of the design.

By using different published options, the contract can be based on traditional competitive tendering, target pricing, or reimbursement of cost. It can even be a management contract. The published options can then be adapted to give further options, for example by introducing a guaranteed maximum price. All these options are available without any change to the core clauses, which include the provisions relevant to claims. The comments that follow on claims under the NEC are therefore relevant to all the published options.

Clarity and simplicity are stated goals of those responsible for the drafting of the NEC, and if the goals are achieved, the need for claims should be substantially reduced. It is probably true that if someone with no previous knowledge of either the NEC or, for example, the ICE Sixth Edition were to read both documents, they would find the NEC much less intimidating. It would certainly take less time to read the NEC.

This apparent clarity may, however, be misleading. The brevity of the published document means that a good deal of supplemental material must be provided by the parties. For example, the contractor may be required to design some or all of the works. The printed clause relating to the contractor's design runs to no more than 17 lines. It says nothing about design standards or responsibilities. These have to be set out in the works information, a document to be prepared by the parties and appended to the contract. Whilst the author of the NEC may have achieved admirable clarity in his document, the effect will be spoiled if the parties are less skilful in preparing documents for the specific project.

The zealous avoidance of traditional phraseology may make the contract more attractive to the lay reader, but many of the old phrases have been the subject of judicial pronouncements by many courts over many years. Their meanings can usually be discovered with the aid of a little research in standard text books. A commentator on the NEC can say what he thinks the contract means, but he cannot be sure until the words in question have been before the courts.

There is no engineer in the NEC. In his place we find two individuals, the supervisor and the project manager. In fact the same individual can fulfil both functions, so long as he remembers in which capacity he is acting at any one time. The supervisor's role is essentially confined to the quality of the work. He carries out whatever tests and inspections may be required and certifies when defects have been made good.

The project manager is responsible for administration, and has very wide duties. There are some 46 specified in the core clauses alone, and more in the option clauses. He is required to deal with all money matters, including claims, but does not have a review function similar to that of the engineer under ICE contracts. If the

contractor disagrees with the project manager the disputes procedure moves straight to adjudication.

The draftsmen of the NEC believed that this new contract was
going to facilitate a completely new approach to contracting, free
from suspicion and confrontation. The first of the core clauses of the
NEC (curiously numbered 10.1) sets the scene for the contract as a
whole:

> The Employer, the Contractor, the Project Manager and the Supervisor
> shall act as stated in this contract and in a spirit of mutual trust and co-
> operation.

Within this one clause we see both the apparent clarity and
simplicity to which the contract aspires, and the difficulties of
uncertainty which it introduces. A contract lawyer assumes that
words are introduced to a contract for a purpose. What is the purpose of this clause? The parties to the contract have an obligation to
'act as stated' without this clause. It is not at all clear whether the
requirement for mutual trust and co-operation adds anything, and
if so, what it adds. The words may well be used by an employer
seeking to avoid a claim made by a contractor, and may also be used
by a contractor seeking to set up a claim. Until these claims and
defences have been raised and litigated, it is not possible to comment authoritatively on their meaning.

It is interesting to note that some employers entering into contracts on the NEC delete clause 10.1. The act of deletion may cause
the contractor to assume that he is not to expect a spirit of mutual
trust and co-operation.

10.3   *Claims under the NEC*

10.3.1   Compensation events

The NEC deals with claims in Section 6 of the core clauses, under
the heading 'Compensation events'. The term also covers other
matters which are not normally considered to be claims, such as
valuation of variations. The term 'compensation event' does not
imply any criticism of the employer or his team – the contractor may
be entitled to time and money for matters which are not in any way
the fault of the employer. In order to be complete the full list is set
out below, but detailed comment is restricted to claims matters.

60.1   The following are compensation events.
(1)   The *Project Manager* gives an instruction changing the Works Information except
   - A change made in order to accept a Defect or
   - A change to the Works Information provided by the *Contractor* for his design which is made at his request or to comply with other Works Information provided by the *Employer*.

The works information will contain

- all the relevant specifications, bills of quantities and drawings provided by or on behalf of the employer;
- statements of any constraints such as required sequences and site access restrictions;
- health and safety requirements;
- details of design to be carried out by the contractor;
- particular requirements with regard to sub-contractors;
- details of bonds and warranties required; and
- a host of other relevant information relevant to the specific project.

If there is an element of contractor's design, there will be some works information supplied by him.

The works information provided by the employer will be listed in Part One of the contract data. That provided by the contractor will be in Part Two.

Works information is not however confined to the materials listed in the contract data. It may be contained in an instruction given under the contract.

The project manager has apparent complete authority to issue any instruction changing the works information, including supplementing it. There is no limit in the contract itself to the possible extent of that change. It seems therefore that the project manager could instruct massive increases in the work, fundamental changes in the nature of the work, or even wholesale omissions. Whether the courts would imply a term that this apparent authority is limited to more conventional 'variations' is not at all clear.

Instructions of the project manager which constitute compensation events therefore include what would normally be classed as variations under conventional contracts, but may go much wider. They certainly include changes in constraints. They do not, however, include a change made in order to accept a defect or a change in works information provided by the contractor (i.e. the contractor's design) made at the contractor's request or to comply with the employer's works information. This does not mean that the project manager cannot issue such instructions. He can certainly do so but such instructions will not be compensation events.

(2) The *Employer* does not give possession of a part of the Site by the later of its *possession date* and the date required by the Accepted Programme.

The employer is required by clause 33.1 to give possession by the later of the possession date, which is defined for relevant sections in Part One of the contract data, and the date required by the accepted programme. If the accepted programme shows a date for possession of part of the site that is earlier than the date given in the contract data, there is no obligation on the employer to meet it. Failure to comply with that clause is a compensation event. It is also a breach of contract. Whether this gives the contractor an alternative remedy is discussed below.

(3) The *Employer* does not provide something which he is to provide by the date for providing it required by the Accepted Programme.

The employer's obligations to provide things for the contractor are difficult to identify in the contract. The contractor can identify dates within the programme when he needs information, materials, or anything else to be provided. The employer, however does not *expressly* undertake to provide such things in accordance with that programme. Nevertheless if he fails to do so that failure will be a compensation event. It may also be a breach of an implied term of contract. If the employer fails to provide information or necessary facilities etc. which are not shown on the accepted programme he may still be in breach of an implied term of contract. This is covered by compensation event (18) below – assuming that the breach can be established.

(4) The *Project Manager* gives an instruction to stop or not to start work.

Any such instruction is a compensation event. It may not carry time or money for the contractor because of the provisions of clause 61.4. (See section 10.3.2 below.)

(5) The Employer or Others do not work within the times shown on the Accepted Programme or do not work within the conditions stated in the Works Information.

Following the conventions of the NEC, 'Others' includes any people or organisations who are not the employer, project manager, supervisor, adjudicator, contractor or contractor's sub-contractor, agent or supplier. 'Others' is not the same as 'other people' to whom reference is made occasionally in the contract. So much for simplicity and clarity.

Failure by anyone to work as required by the programme, including statutory undertakers but not including sub-contractors, gives rise to a compensation event – but only if their activity is shown on the programme. Similarly failure by 'Others' to work within the conditions stated in the works information has the same effect, but only if conditions are set out in the works information. It is therefore essential to the contractor to ensure that any work to be done by someone else which might be critical to his work is shown on the programme or in the works information.

(6) The *Project Manager* or the *Supervisor* does not reply to a communication from the *Contractor* within the period required by this contract.

There are several terms of the contract requiring the project manager or the supervisor to reply to a communication from the contractor within a specific time. Some of the requirements have a time written into them. For example we will find when we discuss the procedures for adding time or money to the compensation event that the project manager may instruct the contractor to submit a quotation. The project manager must reply to that quotation within two weeks. Failure to do may be the cause of further delay or cost and thus is a further compensation event.

There is also a general obligation on the project manager to reply to any communication submitted to him for acceptance (clause 13.4) and if no set time limit relating to the communication concerned is stated in the contract the time is set by the 'period for reply' set out in the contract data.

(7) The *Project Manager* gives an instruction for dealing with an object of value or of historical or other interest found within the Site.

Clause 73.1 obliges the contractor to notify the project manager whenever he finds such an item, and the project manager is required to instruct him how to deal with it. There is no definition of 'other interest' and contractors might be tempted to request an instruction for dealing with matters which may be of relatively little interest. The project manager must be careful how he replies. Unless clause 61.4 applies because there is no effect on actual cost or completion, such an instruction will carry time and money.

> (8) The *Project Manager* or the *Supervisor* changes a decision which he has previously communicated to the *Contractor*.

This is self-explanatory. Again any such change is likely to carry time and money, unless the facts can be brought within clause 61.4.

> (9) The *Project Manager* withholds an acceptance (other than an acceptance of a quotation for acceleration or for not correcting a Defect) for a reason not stated in this contract.

The pressure is on the project manager to respond promptly to all matters needing his decision. This pressure is seen in clause 60.1(6) above and again here. The project manager, for example, is required to accept the programme or give reasons for not accepting it within two weeks (clause 31.3). Any delay in such acceptance will lead to time and money unless the circumstances can be brought within clause 61.4. The project manager must make sure that, if acceptance is withheld, the reason is within the range expressly available to him in the relevant part of the contract.

> (10) The *Supervisor* instructs the *Contractor* to search and no Defect is found unless the search is needed only because the *Contractor* gave insufficient notice of doing work obstructing a required test or inspection.

This seems very favourable to the contractor. If a number of defects have been found, it may be entirely reasonable for the supervisor to instruct a further search in an area where similar defects might be expected. If they are not found, the instruction to search is a compensation event, leading to time and money. But could the employer argue that the instruction 'arises from the fault of the Contractor' and therefore there is no additional payment or time (clause 61.4)? The contract is not at all clear.

(11)   A test or inspection done by the *Supervisor* causes unnecessary delay.

Once again the ability of clause 61.4 to protect the employer from claims is not clear.

(12)   The *Contractor* encounters physical conditions which
   ● are within the Site,
   ● are not weather conditions and
   ● which an experienced contractor would have judged at the Contract Date to have such a small chance of occurring that it would have been unreasonable for him to have allowed for them.

The test here is a little different from the familiar clause 12 in the ICE Conditions. This item exposes, but does not deal with, an uncertainty that has long been recognised in that clause – namely the degree of probability of a physical condition or artificial obstruction that an experienced contractor is required to foresee. The contractor may have realised that it was possible that the physical condition might be encountered, but he may have discounted the possibility. How small the chance has to be before it is discounted will vary from project to project. Presumably all the circumstances of the project, its value, nature, purpose, etc. will have to be taken into account in deciding whether it would have been unreasonable to allow for the small chance of the physical condition arising. An employer may well consider it unreasonable that the contractor should include in his price for the potential cost of dealing with a physical condition that is probably not going to be encountered.
   The weather exclusion is more favourable to the contractor than comparable terms which exclude physical conditions which are *due to* weather conditions. Such a term might exclude flood damage or landslip caused by heavy rain. Those are not weather conditions but may be due to weather conditions. They are not excluded from this clause.

(13)   A *weather measurement* is recorded
   ● within a calendar month,
   ● before the Completion Date for the whole of the *works* and
   ● at the place stated in the Contract Data
   the value of which, by comparison with the *weather data*, is shown to occur on average less frequently than once in ten years.

An attempt is made in this contract to make the assessment of the severity of weather conditions a wholly objective exercise. The standard form contract data states that specific weather measurements are to be taken at a specific place. These can then be compared to previous records of past weather measurements, or if no such records are available a set of constants is provided for comparison purposes. If the actual weather measurement can be shown to occur less frequently than once in ten years, there is a compensation event. This will entitle the contractor to time and, more surprisingly, money.

    (14)   An *Employer's* risk event occurs.

Employer's risk events are defined in clause 80.1. They include matters which have conventionally been described as *force majeure*. If an employer's risk event occurs, the contractor will be entitled to time and money, calculated in accordance with clause 63, which may involve a notional element. On the other hand the contractor is entitled to an indemnity under clause 83.1, which of course will deal only with actual expense incurred. It is not clear how the contractor is expected to split his claim between these inconsistent clauses.

    (15)   The *Project Manager* certifies take over of a part of the *works* before both Completion and the Completion Date.

The certification of early take over (but not take over itself) brings a potential entitlement to time and money. Certification of take over after the completion date, when the contractor is in culpable delay, will bring neither.

    (16)   The *Employer* does not provide materials, facilities and samples for tests as stated in the Works Information.

Again, this is self-explanatory and any such change is likely to carry time and money, unless the facts can be brought within clause 61.4

    (17)   The *Project Manager* notifies a correction to an assumption about the nature of a compensation event.

The need for this provision will be seen when considering the process for assessing time and money consequences of compensation events.

(18)   A breach of contract by the *Employer* which is not one of the other compensation events in this contract.

This is an attempt to ensure that, whatever the Employer and his team do, there is a machinery for extending time and therefore protecting the time provisions of the contract, and the ability of the Employer to recover delay or liquidated damages if Option R is used. It probably does not have the effect of depriving the contractor of his rights to bring claims for damages for breach of contract at common law.

Option B and Option D

If Option B (priced contract with bill of quantities) or Option D (target contract with bill of quantities) is being used, the following sub-clauses are introduced:

60.4   A difference between the final total quantity of work done and the quantity stated for an item in the *bill of quantities* at the Contract Date is a compensation event if
   • the difference causes the Actual Cost per unit of quantity to change and
   • the rate in the *bill of quantities* for the item at the Contract Date multiplied by the final total quantity of work done is more than 0.1% of the total of the Prices at the Contract Date.
   If the Actual Cost per unit of quantity is reduced, the affected rate is reduced.

60.5   A difference between the final total quantity of work done and the quantity for an item stated in the *bill of quantities* at the Contract Date which delays completion is a compensation event.

60.6   The *Project Manager* corrects mistakes in the *bill of quantities* which are departures from the *method of measurement* or are due to ambiguities or inconsistencies. Each such correction is a compensation event which may lead to reduced Prices.

These clauses are essentially valuation matters rather than claims, but they are treated in exactly the same manner as other compensation events.

The Housing Grants, Construction and Regeneration Act 1996

If the contract is a 'construction contract' within the meaning given by section 105 (1) of the Housing Grants, Construction and

Regeneration Act 1996, and was entered into after the Act came into force on 1 May 1998, the parties are likely to have included Option Y(UK)2, which provides:

> 60.7 Suspension of performance is a compensation event if the *Contractor* exercises his right to suspend performance under the Act.

The Act gives the contractor the right to suspend performance of work on notice if he has not been paid in accordance with the contract. That right ceases if and when he is paid. To make the right effective, he must be entitled to an extension of time. This clause does, however, go beyond the requirements of the Act because, by making suspension a compensation event, it entitles the Contractor to a money claim for costs incurred as a result of suspension. Moreover the extension of time assessed under the compensation event procedure may cover the time (and cost) of remobilisation as well as the period of suspension itself.

10.3.2 Notifying compensation events (clause 61)

In order to start the process of assessment of the time and money consequences of a compensation event, one party has to give formal notice to the other of the existence of the compensation event. If the event is one initiated by the project manager or supervisor, the project manager gives notice at the time, and requests a quotation from the contractor. A quotation is not required if the event arises from the fault of the contractor or there has already been a quotation in procedures leading up to the relevant instruction, etc.

It is of course possible that the project manager may fail to request a quotation, either because he does not realise that the instruction he is giving is in fact a compensation event, or perhaps because he would prefer that it should not be seen as such. Moreover many compensation events have nothing to do with the project manager or supervisor at all. The contractor is then required to give notice of the event to the project manager. Time is however critical. The contractor has only two weeks from the date on which he became aware of the event to give his notice. He cannot give notice if more than two weeks has elapsed.

The contractor is not required to submit a quotation when giving his notice of a compensation event and does not do so until asked by the project manager under clause 61.4. That clause provides that if the project manager decides that the event notified by the contractor:

- arises from a fault of the contractor;
- has not happened or is not expected to happen;
- has no effect upon actual cost or completion; or
- is not a compensation event

then there is no change to the prices or the completion date. In other words the basis of the potential claim is rejected, and nothing more needs to be done unless the contractor wishes to dispute the decision. The definition of 'Actual Cost' varies according to the basis of the contract – whether it is a priced or target cost contract, etc. 'Completion' is defined as the date that the contractor in fact finishes his work (clause 11.2(13)).

If on the other hand the project manager agrees that there has been a compensation event, not the fault of the contractor, which has an effect on actual cost and/or completion, he instructs the contractor to submit a quotation. He must do this within one week, unless a longer period has been agreed by the contractor (clause 61.4).

When instructing the contractor to submit a quotation, either because he has taken the view himself that there is a compensation event or because he accepts the contractor's notification of a compensation event, the project manager may advise the contractor that he considers that the contractor has failed in his obligation under clause 16 to give early warning of the event. This is because when the time and money consequences are assessed, they will be assessed as if such early warning had been given (clause 63.4), but only if the project manager has given notice of this when requesting the quotation.

The project manager may be of the view that the effects of the compensation event are too uncertain for a reasonable forecast to be made by the contractor in preparing his quotation. If so, he gives the contractor assumptions on which the contractor is to base his quotation (clause 61.6). If they turn out to be wrong they can be corrected later, such correction being another compensation event (clause 60.1(17)).

Finally, clause 61.7 states that a compensation event is not to be notified after the defects date, the end of the period for correction of defects.

The notification procedure outlined above is designed to ensure that all potential claims are brought out into the open as soon as possible, so that they can be dealt with as they arise and not much later as part of a major and extremely complex claim which is likely to become contentious. The motive in devising this system may be admirable but in practice it can be exceptionally difficult to operate. Each instruction must be the subject of a notice. Indeed any circumstance that may be described as a compensation event must be notified, and the contractor must ensure that he gives notice within two weeks. It can become a practical impossibility to deal with the notices, never mind the quotations that follow, within the time constraints. It is not unusual for the parties to agree at a relatively early stage that the notification procedure and the quotation and assessment provisions that follow are wholly unworkable and should be replaced with some other simpler system. The contractual consequences of this are unpredictable and can be very expensive for one party or the other.

10.3.3 Quotations for compensation events (clause 62)

The project manager may instruct the contractor to submit alternative quotations based on different ways of dealing with the compensation event. The contractor submits the quotations as requested, but may also submit further alternatives based on his own proposed solutions (clause 62.1).

The quotations give proposed changes to the contract prices and the contractor's assessment of delay, with details. Any effect on the programme must be shown by the submission of a revised programme (clause 62.2).

All this has to be done within three weeks. The project manager then has two weeks to reply. He can instruct the preparation of a further quotation, he can accept the quotation already submitted, he can decide not to give the instruction at all (if he had requested the quotation in advance of a proposed instruction), or he can decide to make his own assessment (clause 62.3).

If he decides to request a further quotation he must explain his reasons. The contractor then has a further three weeks (clause 62.4) to supply it.

The time limits can be extended by the project manager with the agreement of the contractor provided the extension is made before time runs out (clause 62.5).

10.3.4   Assessing compensation events (Clause 63)

Compensation events can affect money and time. This is true of all compensation events, including those such as weather related events which traditionally would only affect time.

The money aspect of compensation events is dealt with by changing the prices. 'The Prices' are defined in different ways under the various options but are effectively the contract sum. The change resulting from a compensation event encompasses the effect of the event on the actual cost of the work already done, the forecast actual cost of work not yet done and the resulting fee (if applicable). The term 'Actual Cost' is also defined for each of the options available, and does not mean 'cost actually incurred', but rather 'cost deemed to be actually incurred' in accordance with the provisions of each option. Notional calculations are required, and not simply records of expense.

The fact that it includes 'forecast Actual Cost' adds another important complication. It emphasises that the assessment is not carried out at the end of the job when all the facts are known – it is done more or less at the time of the event itself. If the assessment is wrong, there is no adjustment to update the forecast (clause 65.2). The only adjustment exercise applies to the correction of assumptions stated by the project manager in his instruction to the contractor to submit quotations (clause 61.6) which is itself a compensation event.

The prices are only reduced as a result of a compensation event in the case of a change to the works information, such as an omission of an item, or the correction of an assumption stated by the project manager in his instruction to submit a quotation (clause 63.2).

The words 'extension of time' are not used, but the assessment procedure involves consideration of 'a delay to the Completion Date', which is the same thing. It is assessed as the length of time by which, due to the compensation event, planned completion is later than planned completion as shown on the accepted programme. The delay, or extension of time, is based on the date shown on the programme for completion, not the date in the contract. If the programme shows completion as being ahead of the contractual date, and therefore there is an element of float, the extension of time will be calculated on the basis of the delay to that programmed completion, and then added to the contract date. In other words, the contractor keeps his float.

There is no liquidated damages clause within the basic form of the contract. If the employer wants to include such a clause, he must add in an extra clause. One is provided for him as Option R. The clause is a simple one, stating that the contractor pays delay damages at the rate stated in the contract data from the completion date for each day until the earlier of completion and the date of take over of the works. Any subsequent change to the completion date results in a repayment with interest.

Option R is in fact too simple. There is no procedure for reducing the rate of delay damages if the employer takes over part of the works. Hence, the employer is apparently entitled to take the same amount of damages if 90% if the works have already passed into his possession as he would be entitled to take if none had been handed over. The contractor therefore has a good argument that the figure is a penalty, and therefore unenforceable. The employer should take advice before incorporating Option R, and should consider expanding the standard clause to include a partial possession provision similar to the proviso to clause 47 (1) or (2) of the ICE Sixth Edition.

The assessment of both time and money is affected by the obligation under clause 16 to give early warning of any matter which could increase the prices, delay completion or impair the performance of the works in use. If the project manager concludes that the contractor did not give early warning of a compensation event which an experienced contractor could have given, and gives notice of that view to the contractor, the event is assessed as if he had given such early warning (clause 63.4).

Because the effects of the compensation event are being assessed largely in advance, consideration is also given to the way in which further problems might affect progress in the future as a result of the compensation event. For example, a compensation event that delays progress might push the work into the winter. Weather may then delay matters further, without being serious enough to constitute a further compensation event under clause 60.1(13). Clause 63.5 therefore provides for the assessment of the event to include the cost and time risk for matters which have a significant chance of occurring. There is considerable scope for an imaginative approach but once again the assessment is a once and for all exercise. Assumptions made about problems that are likely to be encountered cannot be corrected later.

Clause 63.6 provides that the assessments are to be made on the assumptions that the contractor will react competently and promptly and that the additional time and cost will be reasonably incurred. It also requires that the assessment will assume that the accepted programme can be changed. If in fact it cannot be changed, perhaps because the work has to be carried out during a weather window, the assessment of the cost and time consequences of the compensation event becomes yet more artificial.

10.3.5 The project manager's assessment (Clause 64)

If the project manager is not satisfied with the contractor's quotation for the time and cost consequences of the compensation event, he makes his own assessment. If he is going to do this he must do so within the time period allowed for the submission of the con-tractor's quotation for the same event, starting from the time when the need for the assessment becomes apparent.

10.3.6 Implementing compensation events (Clause 65)

Clause 65.1 states that the project manager 'implements' or gives effect to each compensation event by accepting the contractor's quotation or completing his own assessment. It becomes effective on such acceptance or assessment, or if that has happened in advance when the event actually occurs. The assessment of the amount due to the contractor at the next assessment date will therefore reflect the changes to the prices resulting from the com-pensation event.

As noted above, clause 65.2 provides that the assessment of the compensation event is not revised if a forecast upon which it is based is shown by later recorded information to have been wrong. The contractor has to stand by his quotation, assuming that it is accepted. He may make a substantial loss or an equally substantial gain, but the employer has the benefit of certainty. If, however, the assessment has been made by the project manager, the contractor may feel seriously aggrieved by a low assessment. He will wish to challenge the assessment, but has limited time in which to do so under the terms of the contract. Whether the terms limit his ability to refer the matter to adjudication under the provisions of the Housing Grants, Construction and Regeneration Act is a moot point discussed below.

Whilst clause 65.2 is applicable to cost reimbursable contracts (Main Options E and F), the effect is of course quite different. The prices (i.e. the amount to be paid) are much more closely related to the real actual cost. The assessment of the compensation event has little to do with the amount being paid.

10.3.7 Challenging the assessment

The New Engineering Contract was one of the first standard forms to have a comprehensive adjudication procedure written into it, and that feature was one of the factors that led Sir Michael Latham to recommend that it be adopted.

The adjudication procedure in the contract did not, however, comply with the requirements of the Housing Grants, Construction and Regeneration Act 1996 and in attempt to bring the contract into line an addendum was published in April 1998. There is, however, real reason to doubt that the attempt was successful.

The procedure under the contract now requires that if the contractor is dissatisfied with an action by the project manager (for example the completion of the assessment of the compensation event) he should give notice no later than four weeks after the action concerned. Within two weeks the contractor and the project manager attend a meeting to discuss and seek to resolve the matter. Clause 90.4 provides that no matter shall be considered to be a dispute unless there has been a notice of dissatisfaction and the matter has not been resolved within four weeks.

There are therefore strict time limits for objection to be taken to an assessment of a compensation event, even though the effects of the compensation event may not be experienced for several months. If notice of dissatisfaction has not been given within four weeks of the assessment, the right to challenge it under the contract is lost. Indeed according to the contract it cannot even be said that there is a dispute.

Whilst this might have been an effective contractual provision if it were able to stand alone, the Housing Grants Construction and Regeneration Act does not permit it to do so. Section 108 requires any construction contract to permit a party to refer any dispute to adjudication at any time. The definition of dispute in the contract cannot restrict the effect of the Act and if the contractor considers that the assessment of the compensation event is incorrect his view is equally able to be the subject of a dispute six months thereafter. He is therefore able to refer the matter to adjudication then.

As with the equivalent amendment to ICE Sixth Edition, this

defect affects the whole adjudication system in the contract. Because the contract does not comply with the Act, the statutory Scheme for Construction Contracts applies, with its own adjudication procedure, which will take precedence over the contractual scheme.

10.4 *Claims for breach of contract at common law*

The principles involved in claims for breach of contract at common law, as opposed to claims for remedies under the contract, have been considered in Chapter 6. Those principles apply to all contracts, whether the standard form in use is one of the ICE family, the New Engineering Contract, or any other.

As has been explained (Chapter 6, section 6.2), a party's right to claim damages for breach of contract can only be excluded by clear words in the contract. Merely providing an alternative remedy, available through the express provisions of the contract, is not enough to exclude the right to claim damages without such clear words.

The compensation event procedure in clause 60 is designed to deal with every possible claim that the contractor may wish to bring, and to deal with such claims as early as possible so as to prevent serious dispute about multiple claim events later on in the contract period. There is, however, no term of the NEC stating that the contractor's rights are limited to claims made through the compensation event procedure. The contractor's common law rights are not excluded.

The difficulties of operating the compensation procedure have been mentioned earlier in this chapter. If contractor misses an opportunity to seek time and money through that procedure he may still be able to bring his claims at common law. This ability is however subject to two important provisos.

Consideration of common law damages will involve asking whether the contractor has properly mitigated his loss. Did he fail to take advantage of the procedure which would have brought him an adequate remedy much earlier? Such failure may have the effect of reducing the damages claimable.

It must also be remembered that not all the listed compensation events are equivalent to breaches of contract. If the project manager or supervisor gives an instruction or otherwise acts as expressly permitted by the contract, there is no breach. For example, if the project manager gives an instruction changing the works information, a compensation event occurs but there is no breach of contract. Similarly the recording of a weather measurement which satisfies the test in clause 60.1(13) cannot be said to be a breach of contract. There is no alternative remedy in such circumstances.

10.5 Sub-contract claims

The comments above regarding the general approach of the NEC apply equally to the standard form of sub-contract. Indeed many of the clauses are identical, save that references are to the parties to the sub-contract instead of the parties to the main contract.

The compensation events are virtually the same as the compensation events in the main contract, discussed earlier in this chapter. The only difference is that whilst various actions or inactions on the part of the employer may give rise to compensation events under the main contract, such actions or inactions can be on the part of either the employer (or his team members, the project manager and/or supervisor) or the contractor in initiating a compensation event under the sub-contract. For example under clause 60.1(5) of the main contract it is a compensation event if the 'Employer or Others do not work within the times shown on the Accepted Programme', but under the sub-contract the equivalent compensation event (with the same clause number) refers to the 'Employer, the Contractor, or Others'.

There is a more significant difference in the procedure for notifying compensation events, which, as under the main contract, is critical to the making of a claim. Whilst the contractor has two weeks to notify a compensation event under the main contract, his sub-contractor only has one week to notify the main contractor. No doubt this is intended to give sufficient time, one week, for the contractor to investigate a sub-contractor's notification and decide whether it is appropriate to give notice under the main contract. It is however an onerous requirement for the sub-contractor. It must also be remembered that in many major projects there will be a chain of sub-contractors. If the main contractor has to notify a compensation event in two weeks, the principal sub-contractor has to notify within one week, the sub-sub-contractors will have little chance of giving notice within the time likely to be given to them.

There are similar problems in the quotation procedure under clause 62. If instructed to produce a quotation, the contractor has three weeks in which to do so but his sub-contractor only has one week. The contractor will therefore have two weeks in which to pull together all the relevant sub-contract quotations before submitting his own. The project manager may take two weeks to reply, and the contractor then has to respond to the sub-contractor.

Whilst the times for submissions are co-ordinated in this way, which may seem acceptable until the effect of multi-level sub-contracting is considered, there is no direct link between compensation events under the main contract and compensation events in the sub-contract. It is quite possible, therefore, that there may be a compensation event under the sub-contract without there being a compensation event under the main contract.

Unlike most traditional sub-contracts it is expressly contemplated that there may be a liquidated damages (or in NEC parlance 'delay damages') provision. If the main contractor is tempted to include Option R he should be aware of the dangers of the standard clause, which are the same as under the main contract outlined above.

CHAPTER ELEVEN
THE CECA FORM OF SUB-CONTRACT

11.1   Introduction

Originally published by the Federation of Civil Engineering Contractors, the 'Form of Sub-Contract for use in conjunction with the ICE Conditions of Contract' is now published by the Civil Engineering Contractors Association, which was founded in November 1996 – the FCEC having been disbanded earlier that year. The current edition is issued in three versions for use in conjunction with the ICE Sixth Edition, the ICE Fifth Edition, and the ICE Design and Construct Contract, dated July 1998, October 1998 and November 1998 respectively. Earlier editions were published by the FCEC at intervals of a few years, generally in order that the form remained compatible with successive revisions to the ICE Conditions.

The principal objective of the form is to enable the main contractor's obligations and rights under the main contract to be reflected, in respect of the sub-contracted works, in corresponding obligations and rights between the sub-contractor and the main contractor. While such an arrangement does not relieve the contractor of his duties under the main contract, it should provide him with indemnity against losses caused by a defaulting sub-contractor whose breach results in the main contractor being in breach of the main contract.

11.2   Sub-contractor's obligations and rights

Clause 2 of the CECA Form provides that:

(1)   The Sub-Contractor shall execute, complete and maintain the Sub-Contract Works in accordance with the Sub-Contract and to the reasonable satisfaction of the Contractor and of the Engineer. The Sub-Contractor shall exercise all reasonable skill care and diligence in

designing any part of the Sub-Contract Works for which design he is responsible.

(2)   The Sub-Contractor shall provide all labour, materials, Sub-Contractor's Equipment, Temporary Works and everything whether of a permanent or temporary nature required for the execution, completion and maintenance of the Sub-Contract Works, except as otherwise agreed in accordance with Clause 4 and set out in the Fourth Schedule hereto.

(3)   The Sub-Contractor shall not assign the whole or any part of the benefit of this Sub-Contract nor shall he sub-let the whole or any part of the Sub-Contract Works without the previous written consent of the Contractor.

  Provided always that the Sub-Contractor may without such consent assign either absolutely or by way of charge any sum which is or may become due and payable to him under this Sub-Contract.

(4)   Copyright of all Drawings, Specifications and the Bill of Quantities (except the pricing thereof) supplied by the Employer or the Engineer or the Contractor shall not pass to the Sub-Contractor but the Sub-Contractor may obtain or make at his own expense any further copies required by him for the purposes of the Sub-Contract. Similarly copyright of all documents supplied by the Sub-Contractor under the Sub-Contract shall remain in the Sub-Contractor but the Employer and the Engineer and the Contractor shall have full power to reproduce and use the same for the purpose of completing, operating, maintaining and adjusting the Works.

'The Sub-Contract Works' is defined in the Agreement as being 'the works which are described in the documents specified in the Second Schedule hereto and which form part of the Works to be executed by the Contractor under the Main Contract'.

  The word 'maintain' used in sub-clause 2(1) above is defined in sub-clause 1(1)(g) as meaning 'the execution of outstanding work and the correction of defects as required by Clause 49 of the Main Contract'. The CECA Form is expressly intended for use in conjunction with the Sixth Edition in which the word 'maintenance' is superseded by the more accurate term 'defects correction', and it is perhaps surprising that the CECA has not adopted the same terminology in clause 2, although it has done so in clause 1.

  Apart from this slight anomaly the provisions of clause 2 are consistent with those of the Sixth Edition.

  In *Costain Civil Engineering Ltd and Tarmac Construction Ltd* v. *Zanen Dredging and Contracting Co Ltd* (1996) the sub-contract was based on a modified version of the September 1984 edition of the FCEC Form. The main contractor (the joint venture) issued a variation instruction to Zanen for work which the arbitrator found, on the facts of the case, not to form part of the main contract. Hence the variation did not constitute a valid instruction under the sub-contract, and the sub-contractor was entitled to payment on a *quantum meruit*. Judge David Wilcox, upholding the arbitrator's award, answered in the affirmative the two questions put to him:

'May the determination of a quantum meruit for such work
(a) be valued by reference to profit allegedly made by the Joint Venture within the sums paid to them for carrying out the work, and/or
(b) be valued by reference to charges ... which a competitor for the said work would have incurred but which Zanen in the circumstances would not have incurred ...?'

The restrictions of the sub-contractor's right to assign benefits of the sub-contract contained in sub-clause 2(3) of the CECA Form, which do not differ materially from those in earlier editions of the FCEC Form, have given rise to several actions in the courts. In *David Charles Flood* v. *Shand Construction Ltd and Others* (1996) it was held in the Court of Appeal, allowing an appeal from the Official Referee, that

'(1) The proviso to clause 2(3) applies only to claims which can be expressed simply as a present or future claim for a fixed amount due under the sub-contract.
(2) Clause 2(3) of the sub-contract rendered invalid the assignment by the company plaintiff of any other claim. It therefore precluded the assignment of claims for damages or for sums which fell to be assessed under or in accordance with the sub-contract terms.'

In their judgment their lordships referred to the previous case of *Yeandle* v. *Wynn Realisations Ltd (in Administration)* (1995) wherein the same issue had arisen a few months earlier and had been determined by the Court of Appeal.

11.3 Main contract provisions

Clause 3 of the CECA form provides that:

(1) The Sub-Contractor shall be deemed to have full knowledge of the provisions of the Main Contract (other than the details of the Con-

tractor's prices thereunder as stated in the Bills of Quantities or Schedules of rates and prices as the case may be), and the Contractor shall, if so requested by the Sub-Contractor, provide the Sub-Contractor with a true copy of the Main Contract (less such details) at the Sub-Contractor's expense. The Main Contractor shall on request provide the Sub-Contractor with a copy of the Appendix to the Form of Tender to the Main Contract together with details of any contract conditions which apply to the Main Contract which differ from the applicable standard ICE Conditions of Contract.

(2)   Save where the provisions of the Sub-Contract otherwise require, the Sub-Contractor shall so execute, complete and maintain the Sub-Contract Works that no act or omission of his in relation thereto shall constitute, cause or contribute to any breach by the Contractor of any of his obligations under the Main Contract and the Sub-Contractor shall, save as aforesaid, assume and perform hereunder all the obligations and liabilities of the Contractor under the Main Contract in relation to the Sub-Contracted Works.

Nothing herein shall be construed as creating any privity of contract between the Sub-Contractor and the Employer.

(3)   The Sub-Contractor shall indemnify the Contractor against every liability which the Contractor may incur to any other person whatsoever and against all claims, demands, proceedings, damages, costs and expenses made against or incurred by the Contractor by reason of any breach by the Sub-Contractor of the Sub-Contract.

(4)   The Sub-Contractor hereby acknowledges that any breach by him of the Sub-Contract may result in the Contractor's committing breaches of and becoming liable in damages under the Main Contract and other contracts made by him in connection with the Main Works and may occasion further loss or expense to the Contractor in connection with the Main Works and all such damages loss and expense are hereby agreed to be within the contemplation of the parties as being probable results of any such breach by the Sub-Contractor.

The usual practice of contractors when inviting tenders from sub-contractors is to send extracts from relevant parts of the bill of quantities and of the specification, with an indication of the applicable ICE or other conditions of contract. Where the standard form of contract is modified, however, details of such modifications are not always provided; and sub-contractors should be careful to ensure that they are aware of any alterations to the standard form that may be detrimental to their interests. A tenderer who, for example, submits a tender for sub-contract work under a main contract which is described as being under 'ICE Sixth Edition (modified)' could find his right to interest on late payments withdrawn by the modifications to the main contract.

Sub-clause 3(4) is clearly intended to ensure that the second rule in *Hadley* v. *Baxendale* (1854) may be applied to losses which do not fall within the category covered by the first rule, namely those that 'arise naturally, according to the usual course of things'. Where for example the value of the main contract is much larger than that of the sub-contract, the amount of liquidated damages specified in the main contract would, if applied to the sub-contract, appear to be disproportionate to its value. Nevertheless delay in completion of the sub-contract works could, if on the critical path, cause delay in completion of the main contract works rendering the contractor liable to the employer for liquidated damages at the rate applicable to the main contract. Relating the sub-contractor's liability to the contractor's liability under the main contract draws the sub-contractor's attention to the potential risk and should generally provide the subcontractor with an incentive to avoid delays that are critical to completion of the main contract works.

11.4   Commencement and completion – delays

Clause 6 of the CECA Form provides that:

(1)   Within 10 days, or such other period as may be agreed in writing, of receipt of the Contractor's written instructions to do so, the Sub-Contractor shall enter upon the Site and commence the execution of the Sub-Contract Works and shall thereafter proceed with the same with due diligence and without any delay, except such as may be expressly sanctioned or ordered by the Contractor or be wholly beyond the control of the Sub-Contractor. Subject to the provisions of this Clause, the Sub-Contractor shall complete the Sub-Contract Works within the Period for Completion specified in the Third Schedule hereto.

(2)   If the Sub-Contractor shall be delayed in the execution of the Sub-Contract Works:

(a)   by any circumstance or occurrence (other than a breach of this Sub-Contract by the Sub-Contractor) entitling the Contractor to an extension of his time for the completion of the Main Works under the Main Contract; or

(b)   by the ordering of any variation of the Sub-Contract Works to which paragraph (a) of this sub-clause does not apply; or

(c)   by any breach of this Sub-Contract by the Contractor

then in any such event the Sub-Contractor shall be entitled to such extension of the Period for Completion as may in all the circumstances be fair and reasonable.

Provided always that in any case to which paragraph (a) of this sub-clause applies it shall be a condition precedent to the Sub-Contractor's right to an extension of the Period for Completion that he shall have given written notice to the Contractor of the circumstances or occurrence which is delaying him within 14 days of such delay first occurring together with full and detailed particulars in justification of the period of extension claimed in order that the claim may be investigated at the time and in any such case the extension shall not in any event exceed the extension of time to which the Contractor is properly entitled under the Main Contract.

(3)   Where differing Periods of Completion are specified in the Third Schedule for different parts of the Sub-Contract Works, then for the purpose of the preceding provisions of this Clause each such part shall be treated separately in accordance with sub-clause (2) above.

(4)   Nothing in this Clause shall be construed as preventing the Sub-Contractor from commencing off the Site any work necessary for the execution of the Sub-Contract Works at any time before receipt of the Contractor's written instructions under sub-clause (1) of this Clause.

(5)   The Contractor shall notify the Sub-Contractor in writing of all extensions of time obtained under the provisions of the Main Contract which affect the Sub-Contract.

In all but the smallest and simplest of sub-contracts it is to be expected that the programme for execution of the sub-contract works will have been discussed and agreed between the sub-contractor and the contractor, and incorporated in the contractor's overall programme for the works, either at the time of preparation of the contractor's clause 14 programme or, in some cases, at the time of preparing tenders. Such programme should be confirmed in writing in order to override the default period of 10 days provided by sub-clause 6(1).

The provision of sub-clause 6(2) above in respect of notices relating to delays originating from causes entitling the main contractor to an extension of time under the main contract should be read in conjunction with the general notice requirements of clause 10(1), under which the sub-contractor's notices must provide the contractor with information needed to satisfy the notice requirements of the main contract. (See section 4.3 above.)

Delays falling within sub-sub-clauses 6(2)((b) and (c) originate from causes for which the main contractor is responsible, and entitle the sub-contractor to 'such extension of the Period for Completion as may in all the circumstances be fair and reasonable'. It remains necessary for the sub-contractor to submit evidence of facts upon which to base a claim for extension of time.

11.5 *Variations*

Clause 8 of the CECA Form states:

(1)   The Sub-Contractor shall make such variations of the Sub-Contract Works, whether by way of addition, modification or omission, as may be:
(a)   ordered by the Engineer under the Main Contract and confirmed in writing to the Sub-Contractor by the Contractor; or
(b)   agreed to be made by the Employer and the Contractor and confirmed in writing to the Sub-Contractor by the Contractor; or
(c)   ordered in writing by the Contractor.
Any order relating to the Sub-Contract Works which is validly given by the Engineer under the Main Contract and constitutes a variation thereunder shall for the purposes of this clause be deemed to constitute a variation of the Sub-Contract Works, if confirmed by the Contractor in accordance with paragraph (a) hereof.
(2)   The Sub-Contractor shall not act upon an unconfirmed order for the variation of the Sub-Contract Works which is directly received by him from the Employer or the Engineer. If the Sub-Contractor shall receive any such direct order, he shall forthwith inform the Contractor's agent or foreman in charge of the Main Works thereof and shall supply him with a copy of such direct order, if given in writing. The Sub-Contractor shall act only upon such order as directed in writing by the Contractor, but the Contractor shall give his directions thereon with all reasonable speed.
(3)   Save as aforesaid the Sub-Contractor shall not make any alteration in or modification of the Sub-Contract Works.
(4)   Variations carried out in accordance with this Clause shall be valued as provided in Clause 9 and payment made in accordance with Clause 15.

Clause 8 of the sub-contract corresponds with clause 51 of the main contract, with the exception that the authority to make variations is vested in the main contractor instead of the engineer. The main contractor has the power to decide whether or not to pass on to the sub-contractor a variation order received from the engineer – and if the engineer by-passes the contractual procedure by giving a direct order to the sub-contractor, the sub-contractor must ignore it except to the extent of advising the main contractor of its existence.

Clause 9 of the CECA form deals with the valuation of variations, and states:

(1)   All authorised variations of the Sub-Contract Works shall be valued in the manner provided by this Clause and the value thereof shall be added to or deducted from the price specified in the Third Schedule hereto or as the case may require.

(2)   The value of all authorised variations shall be ascertained by reference to the rates and prices (if any), specified in the Sub-Contract for the like or analogous work, but if there are no such rates and prices, or if they are not applicable, then such value shall be such as is fair and reasonable in all the circumstances. In determining what is a fair and reasonable valuation, regard shall be had to any valuation made under the Main Contract in respect of the same variation.

(3)   Where an authorised variation of the Sub-Contract Works, which also constitutes an authorised variation under the Main Contract is measured by the Engineer thereunder, then provided that the rates and prices in this Sub-Contract permit such variation to be valued by reference to measurement, the Contractor shall permit the Sub-Contractor to attend any measurement made on behalf of the Engineer and such measurement made under the Main Contract shall also constitute the measurement of the variation for the purposes of this Sub-Contract and it shall be valued accordingly.

(4)   Save where the contrary is expressly stated in any bill of quantities forming part of this Sub-Contract, no quantity stated therein shall be taken to define or limit the extent of any work to be done by the Sub-Contractor in the execution and completion of the Sub-Contract Works, but any difference between the quantity so billed and the actual quantity executed shall be ascertained by measurement, valued under the Clause as if it were an authorised variation and paid in accordance with the provisions of the Sub-Contract.

(5)   Where the Sub-Contractor has been ordered in writing by the Contractor to carry out any additional or substituted work on a daywork basis the Sub-Contractor shall be paid for such work under the conditions set out in the Dayworks Schedule included in the Bill of Quantities under the Main Contract (if any) and at the rates and prices affixed to his Tender or as otherwise agreed. Failing the provision of a Dayworks Schedule the Sub-Contractor shall be paid in accordance with the 'Schedule of Dayworks carried out incidental to Contract Work' issued by the Federation of Civil Engineering Contractors current at the date of the execution of the dayworks.

(6)   Without prejudice to the generality of Clause 3(1) where the Sub-Contractor is to be paid for dayworks at the rates provided in the Dayworks Schedule included in the Bill of Quantities under the Main Contract the Main Contractor shall provide the Sub-Contractor with a copy of the said Dayworks Schedule which should be referred to under Schedule 2.

Clause 9 of the CECA Form corresponds with clause 52 of the ICE Form, and the principles upon which varied work is to be evaluated are similar. (See Chapter 4, section 4.2.)

It is arguable that the provision in the final sentence of sub-clause

9(2), under which valuations of sub-contract work are related to corresponding valuations under the main contract, implies that evidence of such main contract valuations is to be made available to the sub-contractor – and hence that the main contractor is unable to claim privilege in respect of 'without prejudice' correspondence under the main contract.

This issue was considered by Judge Fox-Andrews QC in *Zanen Dredging and Contracting Co Ltd* v. *Costain Civil Engineering Ltd and Tarmac Construction Ltd* (1995). Although the judge held that the words used in sub-clause 10(2) implied that the parties had waived privilege in respect of such correspondence (see next section), he further held, *obiter*, that no such waiver was to be implied in respect of sub-clause 9(2).

11.6 Unforeseeable conditions

Clause 10 of the CECA Form states

(1) Without prejudice to the generality of Clause 3 hereof whenever the Contractor is required by the terms of the Main Contract to give any return, account, notice or other information to the Engineer or to the Employer, the Sub-Contractor shall in relation to the Sub-Contract Works give a similar return, account or notice or such other information in writing to the Contractor as will enable the Contractor to comply with such terms of the Main Contract and shall do so in sufficient time to enable the Contractor to comply with such terms punctually.

Provided always that the Sub-Contractor shall be excused any non-compliance with this sub-clause for so long as he neither knew nor ought to have known of the Contractor's need of any such return, account, notice or information from him.

(2) (a) Subject to the Sub-Contractor's complying with this Sub-Clause 10(2), the Contractor shall take all reasonable steps to secure from the Employer such contractual benefits, if any, as may be claimable in accordance with the Main Contract on account of any adverse physical conditions or artificial obstructions affecting the execution of the Sub-Contract Works and the Sub-Contractor shall in sufficient time afford the Contractor all information and assistance that may be requisite to enable the Contractor to obtain such benefits.

(b) Where the Contractor has claimed additional payment under the Main Contract on account of adverse physical conditions or artificial obstructions affecting the execution of the Sub-Contract Works and the Engineer has determined a sum due to the Contractor by reason of such conditions or obstructions, the Contractor shall within 28 days from such determination by the Engineer determine and notify in writing to the Sub-Contractor such proportion of any such sum which is in all the circumstances fair and reasonable to pay to the Sub-Contractor. Provided that the Contractor shall have no liability to make any such payment to the Sub-Contractor to the extent that the Engineer has determined a sum due to the Contractor by reason of adverse physical conditions or artificial obstructions but the Employer is insolvent and has failed to make payment to the Contractor in respect of such determination.

(c)   Where the Contractor has claimed an extension of time under the Main Contract on account of adverse physical conditions or artificial obstructions affecting the execution of the Sub-Contract Works and the Engineer has determined an extension of time to which the Contractor is entitled by reason of such conditions or obstructions, the Contractor shall determine such proportion of any such extension which is in all the circumstances fair and reasonable to pass on to the Sub-Contractor.

(d)   Save as aforesaid the Contractor shall have no liability to the Sub-Contractor in respect of any condition, obstruction or circumstance that may affect the execution of the Sub-Contract Works and the Sub-Contractor shall be deemed to have satisfied himself as to the correctness and sufficiency of the Price to cover the provision and doing by him of all things necessary for the performance of his obligations under the Sub-Contract. Provided always that nothing in this Clause shall prevent the Sub-Contractor claiming for delays in the execution of the Sub-Contract Works solely by the act or default of the Main Contractor on the ground only that the Main Contractor has no remedy against the Employer for such delay.

(3)   If by reason of any breach by the Sub-Contractor of the provisions of sub-clause (1) of this Clause the Contractor is prevented from recovering any sum from the Employer under the Main Contract in respect of the Main Works, then without prejudice to any other remedy of the Contractor for such breach, the Contractor may deduct such sums from monies otherwise due to the Sub-Contractor under this Sub-Contract.

11.6.1   Disclosing correspondence

Sub-clause 10(2), which in earlier editions of the FCEC Form was expressed more briefly but was substantially similar in its content, gave rise to several appeals from arbitrators' awards, before Official Referees (now judges of the Technology and Construction Court) and in the Court of Appeal. The issues that arose concerned the admissibility in evidence of 'without prejudice' correspondence between the employer, the engineer and the main contractor, and their experts, and the extent of the sub-contractor's entitlement to contractual benefits.

In *Zanen Dredging and Contracting Co Ltd* v. *Costain Civil Engineering Ltd and Tarmac Construction Ltd* (1995) the principal issue related to the admissibility in evidence of certain 'without prejudice' correspondence in the possession of the main contractor (Costain/Tarmac) relating to a dispute under the main contract. Costain/Tarmac relied upon the decision of the House of Lords in *Rush & Tompkins Ltd* v. *Greater London Council & Another* (1988) as authority for their submission that 'without prejudice' correspondence between parties to litigation is protected from production to other parties in the same litigation. Learned counsel for the sub-contractor, Zanen, sought to distinguish the House of Lords' judgment in *Rush & Tompkins* on the ground that sub-clauses 9(2), 10(2) and 15(3)(g) expressly or impliedly waived any 'without prejudice' privilege. Judge Fox-Andrews QC held in the Official Referee's Court that in respect of sub-clause 10(2) it was implicit that, in the event of an issue arising as to whether or not Costain/Tarmac had taken the reasonable steps required under sub-clause 10(2), they would disclose all relevant documents and not object to their admissibility. The judge stated

'If there had been in clause 10(2) of the sub-contract an express provision that in the event of an issue arising as to whether or not Costain/Tarmac had taken the required reasonable steps, they would disclose all documents of the kind listed in paragraph 7 [communications between Costain/Tarmac and the employer, including that of experts on both sides] and not object to their admissibility, it is difficult to see why a court should not enforce such a provision. If a party is prepared to bargain away a right based on public policy to which it would otherwise be entitled, there seems no reason why a court should not hold that party to its contractual obligation. And this would appear to be so whether the privilege is bilateral or unilateral.

There was of course no such express provision in clause 10(2). But I find it was implicit that in the event of such an issue arising a similar obligation was placed upon Costain/Tarmac.'

11.6.2 Contractual benefits

The meaning of the words 'such contractual benefits ... as may be claimable' in clause 10(2)(a) has been considered by the courts in a number of cases; notably *Mooney* v. *Henry Boot Construction Ltd* (1996) and *Balfour Beatty Construction Ltd* v. *Kelston Sparkes Contractors Ltd* (1996), in which appeals from judgments of Judge Humphrey Lloyd QC in the first case and Mr Recorder Crowther QC in the second were heard together in the Court of Appeal.

In the *Mooney* case the judgment of Judge Humphrey Lloyd QC (to the effect that clause 10(2) was merely procedural and that in order to recover under it the sub-contractor had to show an independent entitlement under some other provision of the sub-contract) was overturned. The Court of Appeal held that 'benefit which (1) had been claimed under the main contract in good faith, and (2) has been received by the main contractors, falls to be dealt with in accordance with the second sentence of clause 10(2)'.

In the *Balfour Beatty* case the Court of Appeal held that benefits from instructions under clauses 13 and 51 of the main contract were not benefits under clause 10(2) unless they arose from adverse physical conditions or artificial obstructions, in other words, from clause 12 of the main contract. Furthermore benefits received by the main contractor which contained 'inappropriate elements' were to be treated in the same way provided that they fell within the above definition of 'benefits under clause 10(2)'.

More generally, the Court of Appeal held that 'contractual benefits' are limited to those that are claimable on account of any adverse physical conditions or artificial obstructions which may affect the execution of the sub-contract works. The judgment included the statement

'When there has been an instruction, direction or variation it is the regime of clauses 7, 8 and 9 that applies, and not the somewhat vague assessment of a fair and reasonable proportion under clause 10(2). Or at any rate clauses 7, 8 and 9 operate in preference to clause 10(2) insofar as they deal with the problem, but there may be some residuary matter still to be dealt with under that clause.'

Later in the judgment the question whether 'benefits' under clause 10(2) includes benefits from instructions under clauses 13 and 51 was answered in the negative.

Their lordships appear not to have dealt expressly with the situation where a clause 12 claim by the main contractor results in an instruction under clause 13 or 51 of the main contract, and that instruction is not passed on to the sub-contractor through clauses 7 and 8 of the subcontract. If, as is likely, the clause 12 situation delayed the main contract works and hence also delayed the sub-contract works, the only clause under which the sub-contractor could claim for costs incurred by reason of the delay is clause 10(2). It is submitted that such a circumstance falls within the second sentence of the extract quoted above, as being a 'residuary matter' to be dealt with under clause 10(2), since it is not covered by clauses 7, 8 or 9.

There appears to be a drafting error in sub-clause 10(3) relating to the consequences of a breach by the sub-contractor of the provisions of sub-clause 10(1) which prevents the main contractor from obtaining clause 10(2) benefits. If that situation should arise, there are no benefits to be obtained or to be passed on under sub-clause 10(2). However the sub-clause, construed literally, indicates that the sub-contractor does not merely lose benefits that might have been available: he also suffers a deduction of the sums otherwise due (impliedly sums that would have been recoverable had the sub-contractor complied with sub-clause 10(1)) from monies otherwise due to him.

11.7 Payment

Clause 15 of the CECA Form, which has been modified and expanded substantially from earlier versions in the FCEC Forms, states

> (1) (a) The Sub-Contractor shall not less than 7 days before the date specified in the First Schedule (the 'Specified Date') or otherwise as agreed submit to the Contractor a written statement of the value of all work properly done under the Sub-Contract and of all materials delivered to the Site for incorporation in the Sub-Contract Works and if allowable under the Main Contract the value of off-site materials for incorporation in the Sub-Contract Works at the date of such statement. The statement shall be in such form and contain such details as the Contractor may reasonably require and the value of the work done shall be calculated in accordance with the rates and prices, if any, specified in the Sub-Contract, or if there are no such rates and prices, then by reference to the Price.

(b)   The statement submitted by the Sub-Contractor as provided in the preceding sub-clause shall constitute a 'valid statement' for the purposes of this clause but not otherwise.

(2)   (a)   The Contractor shall make applications for payment in accordance with the Main Contract and subject to the Sub-Contractor having submitted a valid statement shall include in such applications claims for the value of work done and materials set out in such statement.

(b)   In any proceedings instituted by the Contractor against the Employer to enforce payment of monies due under any certificate issued by the Engineer in accordance with the provisions of the main Contract there shall be included all sums certified and unpaid in respect of the Sub-Contract Works.

(3)   (a)   Within 35 days of the Specified Date or otherwise as agreed but subject as hereinafter provided, there shall be due to the Sub-Contractor in respect of the value of the work and materials if included in a valid statement payment of a sum calculated and determined by the Contractor in accordance with the rates and prices specified in this Sub-Contract, or by reference to the Price, as the case may require, but subject to a deduction of previous payments and of retention monies at the rate(s) specified in the Third Schedule hereto until such time as the limit of retention (if any) therein specified has been reached. The Contractor shall notify the Sub-Contractor in writing of the amount so calculated and determined within 35 days of the Specified Date. The final date for payment shall be 3 days later.

(b)   Subject to Clauses 3(4), 10(3) and 17(3) and as hereinafter provided and without prejudice to any rights which exist at Common Law the Contractor shall be entitled to withhold or defer payment of all or part of any sums otherwise due pursuant to the provisions hereof where:–

    (i)   the amounts or quantities included in any valid statement together with any other sums to which the Sub-Contractor might otherwise be entitled do not exceed the amount stated in the Third Schedule, or

    (ii)   the amounts or quantities included in any valid statement together with any sums which are the subject of an application by the Contractor in accordance with Clause 15(2) are insufficient to justify the issue of an interim certificate by the Engineer under the Main Contract, or

    (iii)   the amounts or quantities included in any valid statement are not certified in full by the Engineer, providing such failure to certify is not due to the act or default of the Contractor, or

    (iv)   the Contractor has included the amounts or quantities set out in the valid statement in his own statement in accordance with the Main Contract and the Engineer has certified but the Employer is insolvent and has failed to make payment in full to the Contractor in respect of such amounts or quantities, or

(v) a dispute arises or has arisen between the Sub-Contractor and the Contractor and/or the Contractor and the Employer involving any question of measurement or quantities or any matter included in any such valid statement.

(c) Any payment withheld under the provisions of sub-clauses (b)(iii), (iv) or (v) above shall be limited to the extent that the amounts in any valid statement are not certified, not paid by the Employer or are the subject of a dispute as the case may be.

(d) The provisions of this Clause with regard to the time for payment shall not apply to the amounts or quantities in any valid statement by the Sub-Contractor which are included in the Contractor's statement of final account to the Employer under the provisions of the Main Contract. In respect of any such amounts or quantities the Contractor shall determine and notify in writing to the Sub-Contractor the amount due to the Sub-Contractor, within 28 days of the issue by the Engineer of a certificate stating the amount due to or from the Contractor pursuant to the Main Contract Conditions in respect of the Contractor's statement of final account. Payment of the amount determined by the Contractor shall be due to the Sub-Contractor upon such determination and notification. The final date for payment shall be 7 days later. Provided that the Contractor shall have no liability to make such payment of such amount to the extent that the Engineer has certified such amount or any part thereof pursuant to the provisions of the Main Contract, but the Employer is insolvent and has failed to make payment in full to the Contractor in respect of such certified amount.

(e) In the event of the Contractor failing to make payment of any sum properly due and payable to the Sub-Contractor, the Contractor shall pay to the Sub-Contractor interest on such overdue sum at the rate payable by the Employer to the Contractor under the provisions of the Main Contract. Provided that where the Contractor fails to make payment of any sum properly due and payable to the Sub-Contractor and where the Engineer has certified such amount or any part thereof pursuant to the provisions of the Main Contract, then to the extent that the Employer is insolvent and has failed to make payment to the Contractor in respect of such certified amount, the Contractor shall have no obligation to pay interest to the Sub-Contractor.

(f) Notwithstanding sub-clause (e) the Sub-Contractor shall be paid any interest actually received by the Contractor from the Employer which is attributable to monies due to the Sub-Contractor.

(4) The Contractor shall have power to omit from any determination of the value of work done and materials included in a valid statement the value of any work done, goods or materials supplied or services rendered with which he may for the time being be dissatisfied and for that purpose or for any other reason which to him may seem proper may delete, correct or modify any sum previously determined by him as due for payment to the Sub-Contractor.

(5) (a) Within 35 days of the issue by the Engineer of a Certificate including an amount in respect of payment to the Contractor of the first half of the retention monies or where the Main Works are to be completed by sections for any section in which the Sub-Contract Works are comprised there shall be due to the Sub-Contractor the first half of the retention monies under this Sub-Contract and the Contractor shall so notify the Sub-Contractor in writing. The final date for payment shall be 7 days later. Provided that the Contractor shall have no liability to make such payment if the Employer is insolvent and has failed to release the first half of the retention monies due under the Main Contract.

(b) Within 28 days of the date of issue of the Defects Correction Certificate under the Main Contract, there shall be due to the Sub-Contractor the second half of the retention monies under this Sub-Contract. The Contractor shall so notify the Sub-Contractor in writing within the same 28 day period. The final date for payment shall be 7 days later. Provided that the Contractor shall have no liability to make such payment if the Employer is insolvent and has failed to release the second half of the retention monies due under the Main Contract.

(6) Within three months after the Sub-Contractor has finally performed his obligations under Clause 13 (Outstanding Work and Defects) or within 14 days after the Contractor has recovered full payment under the Main Contract in respect of the Sub-Contract Works, whichever is the sooner and providing that one month has expired since the submission by the Sub-Contractor of his valid statement of final account to the Contractor, the Contractor shall determine the amount finally due under the Sub-Contract from the Contractor to the Sub-Contractor or from the Sub-Contractor to the Contractor as the case may be, after giving credit to the Sub-Contractor for the Price and/or any other sums that may have become due under the Sub-Contract and after giving credit to the Contractor for all amounts previously paid by the Contractor and for all sums to which the Contractor is entitled under the Sub-Contract. The Contractor shall notify the Sub-Contractor in writing

of the amount so determined within the same period. The final date for payment shall be 7 days later.

Provided always that if the Contractor shall have been required by the Main Contract to give to the Employer or to procure the Sub-Contractor to give to the Employer any undertaking as to the completion or maintenance of the Sub-Contract Works, the Sub-Contractor shall not be entitled to payment under this Sub-Contract until he has given a like undertaking to the Contractor or has given the required undertaking to the Employer, as the case may be.

(7) The Contractor shall not be liable to the Sub-Contractor for any matter or thing arising out of or in connection with this Sub-Contract or the carrying out of the Sub-Contract Works unless the Sub-Contractor has made a written claim in respect thereof to the Contractor before the Engineer issues the Defects Correction Certificate in respect of the Main Works, or, where under the Main Contract the Main Works are to be completed by sections, the Defects Correction Certificate in respect of the last of such sections in which the Sub-Contract Works are comprised.

(8) Every written notification given by the Contractor to the Sub-Contractor of amounts of payments due to the Sub-Contractor under this Sub-Contract shall specify the amount (if any) of the payment made or proposed to be made and the basis on which that amount was calculated.

(9) In the event of the Contractor withholding any payment after any final date for payment hereunder, he shall notify the Sub-Contractor of his reasons in writing not less than one day before the final date for payment specifying the amount proposed to be withheld and the ground for withholding payment or if there is more than one ground each ground and the amount attributable to it.

The three versions of the CECA Form, which were all published within the second half of 1998 have not, as at April 1999, given rise to direct authorities, but much of the case law relating to earlier (FCEC) forms remains relevant to the CECA forms. In the case of the payment provisions, however, clause 15 has been substantially revised in the CECA form, principally in order to give greater protection to the sub-contractor against unjustified withholding by the main contractor of payments due him. Where sums paid to the sub-contractor are less than those included in his statement, the contractor is required by the CECA form to notify the sub-contractor of the sum calculated as being payable, and to provide details of any payments withheld by reason of

- deductions made by the engineer;
- failure of the employer because of his insolvency to make the full payment due; or
- a dispute under either the main contract or the sub-contract.

In *Zanen Dredging and Contracting Co Ltd* v. *Costain Civil Engineering Ltd and Tarmac Construction Ltd* (1995) (see section 11.6 above) a question arose as to whether sub-clause 15(3)(g) of the 1984 revision (which reappears as sub-clause 15(3)(f) of the CECA form) implied waiver of privilege in respect of 'without prejudice' correspondence between the main contractor and the employer relevant to the matters in issue. Judge Fox-Andrews QC held in the Official Referee's Court (as it then was) that there was no reason to imply into sub-clause 15(3)(g) a term waiving privilege, as he had done in the case of sub-clause 10(2).

11.8   Disputes

Clause 18 of the CECA form has been greatly expanded from the clause 18 that appeared in earlier versions of the FCEC form, mainly by the inclusion of procedures for dealing with disputes by informal discussions, conciliation or adjudication.

11.8.1   Conciliation

Conciliation depends for its success upon a genuine wish by both parties to achieve a settlement at the earliest possible date; and, while the ICE Conciliation Procedure recognises that aim by prescribing limited periods for the various stages of the procedure, it may still cause unnecessary delay where one party wishes to use it solely for the purpose of deferring the date on which payment has to be made. The outcome of a reference to conciliation is a recommendation, which may be rejected by either party within 28 days of its publication by their serving a Notice of Adjudication or a Notice to Refer to Arbitration. Where this happens, the time spent and the cost incurred in the conciliation will have been wasted.

11.8.2   Adjudication

The Housing Grants, Construction and Regeneration Act 1996 ('the Act'), which was enacted on 24 July 1996, includes in Part II provisions relating to construction contracts falling within a definition of that term contained in section 104 of the Act. Section 114 of the Act empowered the Secretary of State to publish the *Scheme for*

Construction Contracts ('the Scheme'), which scheme was published in March 1998 and came into force on 1 May 1998.

The Scheme provides, in regulation 2, that where a construction contract 'does not comply with the requirements of section 108(1) to (4) of the Act, the adjudication provisions in Part I of the Schedule to these Regulations shall apply'. That section of the Act gives the parties to a construction contract the right to refer a dispute to adjudication, defines the machinery by which the adjudication is to be implemented, and in subsection (3) states that

'The contract shall provide that the decision of the adjudicator is binding until the dispute is finally determined by legal proceedings, by arbitration (if the contract provides for arbitration or the parties otherwise agree to arbitration) or by agreement.

The parties may agree to accept the decision of the adjudicator as finally determining the dispute.'

In *Macob Civil Engineering Ltd* v. *Morrison Construction Ltd* (1999) the sub-contractor ('Macob') sought enforcement of an adjudicator's decision awarding it some £300 000 plus VAT and interest against the main contractor (Morrison). It was common ground that the contract was a construction contract within the meaning of the Act and that the contract did not comply with all of the requirements of section 108(1) to (4) of the Act so that the provisions of the Scheme applied. Counsel for Morrison challenged the adjudicator's decision on two grounds alleging breaches of the rules of natural justice, and in addition on the ground that the adjudicator had no power to make a peremptory decision which he had purported to do.

Giving judgment in the Technology and Construction Court for Macob Mr Justice Dyson commented that it was the first time that the court had had to consider the adjudication provisions of the Housing Grants, Construction and Regeneration Act 1996. The judge quoted certain provisions of Part I of the Scheme as being the only material provisions to which he need refer, as follows

'23(1) In his decision the adjudicator may, if he thinks fit, order any of the parties to comply peremptorily with his decision or any part of it.

(2) The decision of an adjudicator shall be binding on the parties, and they shall comply with it until the dispute is finally determined by legal proceedings, by arbitration (if the contract

provides for arbitration or the parties otherwise agree to arbitration) or by agreement between the parties.'

He also quoted section 42 of the Arbitration Act 1996, as modified in accordance with regulation 24 of the Scheme:

'(1) Unless otherwise agreed by the parties, the court may make an order requiring a party to comply with a peremptory order made by an adjudicator.'
(2) An application for an order under this section may be made . . .
(b) by a party to adjudication with the permission of the adjudicator (and upon notice to the other parties).'

The judge accepted the submission of counsel for Macob, that 'even if there is a challenge to the validity of an adjudicator's decision, the decision is binding and enforceable until the challenge is finally determined'. He commented:

'The intention of Parliament in enacting the Act was plain. It was to introduce a speedy mechanism for settling disputes in construction contracts on a provisional interim basis, and requiring the decisions of adjudicators to be enforced pending final determination by arbitration, litigation or agreement: see section 108(3) of the Act and paragraph 23(2) of Part I of the Scheme . . .

It is clear that Parliament intended that the adjudication should be conducted in a manner which those familiar with the grinding detail of the traditional approach to the resolution of construction disputes apparently find difficult to accept. But Parliament has not abolished arbitration and litigation of construction disputes. It has merely introduced an intervening provisional stage in the dispute resolution process. Crucially, it has made it clear that decisions of adjudicators are binding and are to be complied with until the dispute is finally resolved.

It is well known that many, if not most, construction contracts contain arbitration clauses. It is by no means uncommon for such clauses in sub-contracts to state that the arbitration between main contractor and sub-contractor may not be commenced until the main contract works have been completed, at any rate unless the main contractor decides otherwise. The sub-contract in the present case provides an example of this. In such a case, the groundworks sub-contractor to a major development may have to wait years before he can even start to arbitrate his dispute with the main contractor. This was the mischief at which the Act was aimed.'

The decision of an adjudicator is binding unless and until overturned in arbitration or other dispute resolution procedure adopted by the parties. Hence in cases where the sub-contractor is not receiving the payments to which he considers himself to be entitled, adjudication may provide at least a temporary alleviation of his financial difficulties.

11.8.3 Arbitration

Where a dispute is not referred to conciliation or adjudication, or where the outcome of such procedures is unacceptable to either party, the dispute is determined by arbitration or by such other means as the parties may elect. Sub-clause 18(7) provides:

(a) All disputes arising under or in connection with the Sub-Contract, other than failure to give effect to a decision of an adjudicator, shall be finally determined by reference to arbitration. The party seeking arbitration shall serve on the other a notice in writing (called the Notice to Refer) to refer the dispute to arbitration...'

Sub-clause 18(8) provides:

(a) The arbitrator shall be a person appointed by agreement of the parties.
(b) If the parties fail to appoint an arbitrator within 28 days of either party serving on the other party a notice in writing (hereinafter called the Notice to Concur) to concur in the appointment of an arbitrator the dispute shall be referred to a person to be appointed on the application of either party by the President for the time being of the Institution of Civil Engineers.

Sub-clause 18(10) provides, *inter alia*:

(c) If the Contractor is of the opinion that a Main Contract Dispute has any connection with a dispute in connection with the Sub-Contract (hereinafter called a Related Sub-Contract Dispute) and the Main Contract Dispute is referred to an arbitrator under the Main Contract, the Contractor may by notice in writing require that the Sub-Contractor provide such information and attend such meetings in connection with the Main Contract Dispute as the Contractor may request. The Contractor may also by notice in writing require that any Related Sub-Contract Dispute be dealt with jointly with the Main Contract Dispute and in a like manner. In connection with any Related Sub-Contract Dispute the Sub-Contractor shall be bound in like manner as the Contractor by any award by an arbitrator in relation to the Main Contract Dispute.

(d) If a dispute arises under or in connection with the Sub-Contract (hereinafter called a Sub-Contract Dispute) and the Contractor is of the opinion that the Sub-Contract Dispute raises a matter or has any connection with a matter which the Contractor wishes to refer to arbitration under the Main Contract, the Contractor may by notice in writing require that the Sub-Contract Dispute be finally determined jointly with any arbitration to be commenced in accordance with the Main Contract. In connection with the Sub-Contract Dispute, the Sub-Contractor shall be bound in like manner as the Contractor by any award of an arbitrator concerning the matter referred to arbitration under the Main Contract.

Sub-clause 18(10)(d) appeared with slightly different wording as sub-clause 18(2) of the September 1984 revision of the FCEC form and as sub-clause 18(8) of the September 1991 revision.

Much of the case law arising from clause 18 of the FCEC Form relates to the provision made for consolidation of arbitrations under the sub-contract with those arising under the main contract. In *Wynn Realisations Ltd (in Administration)* v. *Geoff Yeandle Contractors Ltd* (1992) the main contract was under the ICE Fifth Edition and the sub-contract under the September 1984 revision of the FCEC Form. It was held by Judge Potter in the Official Referee's Court that a dispute had arisen under the main contract which related to the sub-contract works and that a notice by the main contractor under clause 18(2) was effective in rendering invalid the arbitrator's appointment (by the ICE) under the sub-contract, and hence that arbitrator was without jurisdiction. There was, however, no obligation upon the main contractor to proceed with the main contract arbitration. Therefore the sub-contractor had no means of invoking the arbitration clause until such time as the main contractor chose to exercise his right under clause 66 of the main contract.

That decision was effectively overturned in *Erith Contractors Ltd* v. *Costain Civil Engineering Ltd* (1993) in which Judge John Loyd QC held that

'It is in my view axiomatic that if the contractor requires the dispute to be dealt with jointly with his dispute with the employer in accordance with the provisions of clause 66 of the main contract, he is under an obligation to take the necessary steps to have the dispute dealt with in accordance with the provisions of clause 66.'

In the Scottish case of *Dew Group Ltd* v. *Costain Building and Civil Engineering Ltd* (1996) a sub-contract between the parties was based on the September 1984 revision of the FCEC Form of Subcontract. The pursuers, Dew, gave notice to the defenders, Costain, referring disputes that had arisen to arbitration, and proposing the appointment of one of two named persons as arbiter. Costain replied rejecting the two persons proposed by Dew because they were not on the ICE list of arbitrators, and proposing either of two other named persons. Dew agreed to the appointment of one of those persons, Ian Menzies, 'provided he was willing and able to act'. Dew then wrote to Mr Menzies inviting him to accept the appointment and to provide his terms and conditions of appointment. Mr Menzies replied accepting the appointment, specifying his terms, but advising the parties that because of other commitments he would be unable to hold a substantive hearing within the coming twelve months.

Wishing to avoid the delay, Dew wrote to Costain suggesting that the other person named by Costain be contacted. On the same day Costain wrote to Dew saying that it had commenced proceedings under the main contract in the Court of Session and giving notice under clause 18(3) of the sub-contract, which provided (in the 1984 form of sub-contract), for consolidation of disputes arising from the main and the sub-contract where notice is given *before* an arbitrator is appointed pursuant to sub-clause 18(1).

It was held in the Court of Session, Outer House, that Dew's letter agreeing to the appointment of Mr Menzies constituted a conclusive agreement as to a named person to act as arbiter, and therefore that the arbitration should proceed.

More recently, in *Redland Aggregates Ltd* v. *Shepherd Hill Engineering Ltd* (1998), it was held in the Court of Appeal that clause 18(2) provides for a tripartite arbitration. It was, however, recognised by the court that '[i]t may well be that, on the true interpretation of the main contract, the employer cannot be compelled to participate in a tripartite arbitration'. Secondly the Court of Appeal quoted with approval the words of Mr Justice Steyn (as he then was) in *M J Gleeson Group plc* v. *Wyatt of Snetterton Ltd* (1994):

'I have no hesitation in holding that there is implied in [clause 18(2)] a term that, if the contractors are unable or unwilling to bring about a tripartite arbitration within the time span contemplated by the clause, the sub-contractors will be free to proceed with an arbitration under clause 18(1). That seems to me necessary to give business efficacy to the contract.'

Allowing the appeal the Court of Appeal declared that 'the plaintiffs are no longer obliged to take part in a tripartite arbitration, and may call upon the President of the Institution of Civil Engineers to appoint an arbitrator on their disputes with the defendants'.

11.8.4 Conclusion

Although a relatively simple form in comparison with other standard forms of construction contract the CECA form of sub-contract and, more particularly, its predecessors the FCEC forms have given rise to many disputes. It remains to be seen whether the improvements introduced in the latter half of the 1990s – especially the 1996 Arbitration Act and the 1996 Construction Act – will be effective in leading to more rapid and economical dispute resolution procedures.

TABLE OF CASES

TABLE OF CONTRACT CLAUSES

INDEX